191-75

Plasticity refers to the ability of many organisms to change their biology or behaviour to respond to changes in the environment, particularly when these are stressful. Humans are, perhaps, the most plastic of all species, and hence the most variable.

This book reflects on the history of research in this area, state-of-the-art research methods and discoveries, and needs for future research in human plasticity and variability.

Topics discussed include child growth, starvation, disease of both young and old, and the effects of migration, modernization and other life-style changes.

The book will be especially useful to biological and physical anthropologists, human biologists and medical scientists interested in knowing more about how and why humans vary.

Cambridge Studies in Biological Anthropology 15

Human variability and plasticity

Cambridge Studies in Biological Anthropology

Series Editors

G.W. Lasker
Department of Anatomy and Cell Biology,
Wayne State University,
Detroit, Michigan, USA

C.G.N. Mascie-Taylor
Department of Biological Anthropology,
University of Cambridge

D.F. Roberts
Department of Human Genetics,
University of Newcastle-upon-Tyne

R.A. Foley
Department of Biological Anthropology,
University of Cambridge

Also in the series

Barry Bogin   *Patterns of Human Growth*

Julius A. Kieser   *Human Adult Odontometrics*

J.E. Lindsay Carter and Barbara Honeyman Heath   *Somatotyping*

Roy J. Shephard   *Body Composition in Biological Anthropology*

Ashley H. Robins   *Biological Perspectives on Human Pigmentation*

C.G.N. Mascie-Taylor and G.W. Lasker (editors)   *Applications of Biological Anthropology to Human Affairs*

Alex F. Roche   *Growth, Maturation, and Body Composition*

Eric J. Devor (editor)   *Molecular Applications in Biological Anthropology*

Kenneth M. Weiss   *The Genetic Causes of Human Disease*

Duane Quiatt and Vernon Reynolds   *Primate Behaviour*

Stanley J. Ulijaszek and C.G.N. Mascie-Taylor (editors)   *Anthropometry: The Individual and the Population*

Stanley J. Ulijaszek   *Human Energetics in Biological Anthropology*

Human variability and plasticity

EDITED BY

C.G.N. MASCIE-TAYLOR
Department of Biological Anthropology,
University of Cambridge,
Cambridge, UK

AND

BARRY BOGIN
Department of Behavioral Science,
University of Michigan,
Dearborn, Michigan, USA

CAMBRIDGE
UNIVERSITY PRESS

Published by the Press Syndicate of the University of Cambridge
The Pitt Building, Trumpington Street, Cambridge CB2 1RP
40 West 20th Street, New York, NY 10011-4211, USA
10 Stamford Road, Oakleigh, Melbourne 3166, Australia

First published 1995

Printed in Great Britain at the University Press, Cambridge

A catalogue record for this book is available from the British Library

Library of Congress cataloguing in publication data
Human variability and plasticity / edited by C.G.N. Mascie-Taylor and
B. Bogin.
    p.   cm. – (Cambridge studies in biological anthropology : 15)
Includes index.
ISBN 0 521 45399 2
1. Physical anthropology. 2. Human anatomy – Variation.
3. Adaptation (Physiology). 4. Biological diversity. 5. Man – Influence of environment. I. Mascie-Taylor, C.G.N. II. Bogin, Barry. III.
Series.
GN62.8.H87 1995
573 – dc20 94-33686 CIP

ISBN 0 521 45399 2 hardback

VN

This volume is published
to honour
the professional life of

GABRIEL W. LASKER

Contents

ix

Contributors

Professor B. Bogin
Department of Behavioral Sciences, The University of Michigan, Dearborn MI 48128, USA

Dr J.L. Boldsen
Department of Social Medicine, Odense Universiteit, J.B. Winsløws Vej 17, DK-5000, Odense C, Denmark

Dr D. Coleman
Department of Applied Social Studies and Social Research, University of Oxford, Wellington Square, Oxford OX1 2ER, UK

Dr R.M. Garruto
National Institute of Health, Building 36, Room 5B21, Bethseda, MD 20892, USA

Professor G.A. Harrison
Institute of Biological Anthropology, University of Oxford, 58 Banbury Road, Oxford OX2 6QS, UK

Professor G.W. Lasker
Department of Anatomy and Cell Biology, School of Medicine, Wayne State University, Detroit, MI 48201, USA

Dr C.G.N. Mascie-Taylor
Department of Biological Anthropology, University of Cambridge, Downing Street, Cambridge CB2 3DZ, UK

Dr G.E.H. Mohamed
Department of Biological Anthropology, University of Cambridge, Downing Street, Cambridge CB2 3DZ, UK

Dr D.J. Pritchard
Department of Human Genetics, University of Newcastle Upon Tyne, 19 Claremont Place, Newcastle upon Tyne, NE2 4AA, UK

Dr J.H. Relethford
Department of Anthropology, State University of New York, Oneonta, NY 13920-4015, USA

Professor D.F. Roberts
Department of Human Genetics, University of Newcastle upon Tyne, 19 Claremont Place, Newcastle upon Tyne NE2 4AA, UK

Professor L.M. Schell
Department of Anthropology, State University of New York, Albany, NY 12222, USA

Dr S.J. Ulijaszek
Department of Biological Anthropology, University of Cambridge, Downing Street, Cambridge CB2 3DZ, UK

Foreword

Physical anthropologists and even their successors, biological anthropologists, have often perceived the influence of environmental variation on developmental processes as a dreadful nuisance. This, of course, it is if one is using the phenotypes as measures of the genotypes that are required for establishing affinities and evolutionary relationships between populations; and such objectives dominated anthropology throughout most of this century. When, however, attention is turned to questions of developmental regulation and growth, adaptability and homeostasis, fitness and health, the environmental factors that affect these processes and phenomena become the essential focus of interest. In many respects (though not all) it is the genetic variance that becomes the nuisance. Increasingly, such issues have become of concern in biological anthropology, and the ecological dimension of the discipline is now considered to be as important as the evolutionary one. Ultimately, anyway, they represent different aspects of much the same coin.

The recognition of the importance of environmental dimensions owes much to the researches of Gabriel Lasker. At the time that he was undertaking his pioneering studies in Mexico, almost nothing else was being done on the nature of plasticity in humans, only in animals. He recognized not only the importance of the phenomenon but also how, in the much more difficult circumstances presented in human biology, it could be rigorously investigated by natural experiments, especially those involving migration.

The contributions to this book by colleagues and friends of Gabriel Lasker pay tribute to that pioneering work and well show the range of enquiries opened up by it. The concept of plasticity is widened well beyond the developmental one with, for example, contributions on demography, isonymy and epidemiology, but these are all fields in which Gabriel Lasker has made notable achievements. The breadth of his own vision in human population biology is without limit. And isn't it both fitting and significant that he is also an author in this volume? Instead of just glorying in the praise he is, as always, out in the front line fighting with the troops, and no doubt attempting to ensure that they get it right! Age has certainly been no

xiii

handicap for Gabriel, who seems, if anything, to gather enthusiasm and productivity with increasing years.

Geoffrey Harrison

1 *The pervasiveness of plasticity*

D. F. ROBERTS

Summary

It is shown that plasticity is one of the key concepts in human biology; it provides a link between the widely differing areas of the subject in which it is manifested, and so gives a unity of approach and understanding that holds the subject together. Earlier studies tended to be empirical, concerned mainly with body size and form. Knowledge was refined by experimental studies, some exploiting 'natural experiments', situations that have occurred without scientific planning; some monitoring the effects of changes introduced for purposes other than experiment; others examining the effects of chosen variables in planned experimental situations. It is this last type of study that has helped us to understand the mechanisms of the plasticity response. As a result, the pervasiveness of plasticity is now apparent, its effects being discernible in features ranging from gross variables such as body size down to details of body composition and the distribution and structure of tissues. It occurs not only during the growing period but also in adulthood and senescence. The occurrence of plasticity brought about by functional adaptation of the individual, supported by genetic experiment, suggests an evolutionary process additional to natural selection by which organisms achieve long-term adaptation to their environments.

Introduction

'Plasticity' derives from the family of Greek words with the root *plassu*, e.g. *plasticos*, and is defined in the Oxford English Dictionary as 'the capability of being moulded'.

I first came across the word applied in biology by Huxley (1942), who used it with several connotations. It denoted: alternative genic expressions, operated by switch mechanisms to give a few contrasted phenotypes; the evolutionary advantage conferred by sexual reproduction and especially the potential for recombination of genes; the evolutionary advantage conferred by diploidy which allows the species to carry a store of recessives unexposed to selection; the advantage possessed by a species that is subdivided into subpopulations with their harmoniously stabilized gene

1

complexes, as set out by Sewall Wright. Then Dobzhansky (1957) noted that a species becomes adapted in two ways, involving on the one hand genetic specialization and on the other adaptive plasticity of the phenotype. In the first it evolves a variety of genotypes each specialized to fit part of the available range of environments. In the second it evolves genotypes that permit their possessors to adjust themselves to a spectrum of environments by homeostatic modification of the phenotype. The word, and indeed the concept, appeared late in physical anthropology, partly on account of the typological approach of the earlier decades. In Harrison *et al.*'s text *Human Biology* the word did not appear until the third edition in 1987. I first came across it as applied to humans in 1961 in the little book by Lasker, *The Evolution of Man*. There he used it to refer to 'capacity of the individual to change in response to his environment' (p. 206) and qualified this (p. 208) as 'capacity to change within the lifetime of the individual. It applies especially to those permanent effects that may occur as a result of changed environment during the growth period'. In the glossary to the second edition (1976) the definition is formalized as 'ability of the individual to be adaptively modified by response to the environment during growth'.

In Lasker's 1969 paper he saw plasticity as one of three levels of human adaptation. There is natural selection of genotypes, directly affecting the genetic constitution of the population. Secondly, there is developmental or ontogenetic adaptation, which has no genetic effect other than to reduce the necessity for adaptive natural selection. Thirdly, there are physiological and behavioural responses, changes that are reversible, adapting individuals to the immediate environment.

The first investigations into plasticity concerned its more restricted sense, enquiring whether body size and form were subject to any appreciable modification under varying circumstances. At a time when racial types were generally assumed to be fixed, even this definition of plasticity was only gradually demonstrated; in fact, most of Lasker's own work deals with this point rather than with the adaptive nature of plastic changes. There is now a considerable body of evidence on the extent and nature of plasticity in humans, deriving from studies of different kinds.

Empirical studies

A number of early empirical studies showed that immediate environment affects the body form. In Europe Walcher (1905, 1911) showed that when babies are habitually placed supine, so that the weight of the head falls on the occiput, they become broader-headed than do babies placed so that they lie on their side. The same occurs in some cradling practices where the babies are fastened into cradles in such a way that the back of the head is pressed constantly upon the hard wooden headboard (Ehrich & Coon, 1947). In response to the still more severe environmental stress of artificial cranial deformation, the plasticity of the skull is shown not only by the altered contour and angular relationships but also by altered probabilities of occurrence of minor morphological variants such as Wormian bones (Ossenberg, 1970). The study of Brezina & Wastl (1929) in Vienna showed that hand dimensions were related to the habitual activities practised in the course of a man's occupation. A remarkable study was that of Ivanovski (1925) who undertook periodic investigations with colleagues in different parts of Russia to trace the physical effects of the Russian famine. Altogether 2114 individuals were measured, and each individual was measured six times during the course of the investigation. In every regional subsample there were diminutions in stature, head length, head breadth, head circumference, face length and breadth, stem length and relative chest girth; many other characters were also affected.

Most of the early human evidence usually cited for plasticity comes from the remarkable series of studies from the United States prompted, as Lasker pointed out, by the considerable interest at that time in the assimilation of immigrants. The first evidence was incidental. During the civil war, Gould (an astronomer) analysed records on over a million recruits for the northern armies. Among his findings was the demonstration that the American-born were taller than the European-born. Unfortunately these data included many subjects where the height was self-reported or had been guessed at, but the same conclusion emerged from Baxter's (1875) reanalysis of the data omitting all records where the measurements had not been taken by standardized procedures by medical officers. Bowditch (1879) showed that children of non-labouring classes in Boston tended to be taller than those in the labouring classes, but the difference was considerably less than in comparable classes of boys in England. A more direct study was that of Fishberg (1905) demonstrating differences in stature and cephalic index between Jewish groups in the United States and in Europe. Boas interpreted the findings of Gould, Baxter and Bowditch as indicating that under more favourable environments the physical development of a race may improve. In 1910 he commenced a survey of the physical measurements of immigrants to the United States and their families and found that, by comparison with the children born in Europe before the parents migrated, the children born after their arrival in America were larger and of different head shape. As Tanner (1981) so succinctly put it, 'his study was vast in scope, unexpected in outcome, and staggering in professional and public reception'. Boas concluded (1912) 'the adaptability of the immigrants seems to be very much greater than we had a right to suppose before our investigations were instituted' and that the changes were a result of some aspect of the American environment. These results caused considerable controversy, for they challenged the whole concept of the fixity of type, and in particular head shape, that was prevalent in anthropology at that time.

One possible explanation of some of the immigrant difference might be that migrants differed in some respects from those who stayed at home, as was subsequently shown in Martin's study (1949) of height, weight, and chest girth in some 91 000 young men aged 20–21 in Britain, and by Mascie-Taylor (1984) especially for height. Shapiro (1939) initiated a study to enquire whether self-selection of migrants contributed to the differences. He examined Japanese born in Hawaii, Japanese migrants to Hawaii, and Japanese who had stayed in Japan. Not only did the Japanese born in Hawaii differ from the new migrants in many features including stature and cephalic index, but the migrants also differed from those who had remained at home in Japan. Shapiro attributed the latter difference to self-selection, and the difference between Hawaiian-born and migrants to an environmental effect during the growth period.

There are other possible confounding factors that subsequent studies took into account. Secular increase in stature is a widespread phenomenon, which may have entered into the comparisons of migrants, sedentes and local-born. Goldstein (1943) examined comparable generations, samples in Mexico of Mexican parents and their adult children and, in the United States, of Mexican immigrants and their adult American-born children. As now was to be expected, the immigrants were larger than those who had remained in Mexico, the younger adults in Mexico were larger than their parents, and those in the United States were larger than their parents. The fact that this last difference was the most pronounced suggested some influence on growth that was especially prevalent in the United States. Relevant to this interpretation is the work of Lasker (1952, 1954) who compared measurements in Mexicans who had migrated as children or as adults to the United States with the corresponding measurements in Mexican sedentes; he also compared brothers who migrated at different ages or who did not migrate. There was no evidence for any self-selection of migrants in terms of size. Nor was there in the work of Malina *et al.* (1982) who followed up migrants from Oaxaca to Mexico City, measured before they had moved, and found no differences from those who remained at home.

These several studies all consisted of comparisons of samples of populations. There may of course have been genetic differences between the samples. Hulse (1968) compared people of Ticino canton in Switzerland with their close relatives in California. He used the frequency of eye colours to show that the two groups were comparable genetically. All the subjects could be regarded as members of the same breeding population. In Switzerland there was a strong tendency to choosing a spouse not only from the same canton but especially from the same village (village endogamy); in California, although the preference was strongly for a spouse from Ticino, it was difficult to find one from the same ancestral village or town, so that the Californians included more individuals who were offspring of exogamous marriages. Hulse took exogamy into account in his analysis, and showed that not only were the Californian-born taller and longer-headed than those born in Switzerland, but in both areas these differences characterized offspring of exogamous compared with endogamous unions. Again Hulse partitioned the causes of these differences. He attributed the greater stature, and therefore the lower cephalic index that is correlated with it, to the superior nutrition of the Californian-born. The differences in both regions between those of exogamous and endogamous parentage he attributed to heterosis.

There were other studies of migrants of ancestral origins other than European. For example, heights and weights of 300 children born and brought up in Canada, with Punjabi parents, were compared with those of Punjabi children born in India. The Canadian-born children were taller and much heavier at all age groups than those born and brought up in India (Someswara Rao *et al.*, 1954). Greulich (1957) compared Japanese children reared in California with those reared in the then poorer environment of Japan. The California-reared children were bigger at all ages, although there was no change in the relationship of sitting height to leg length. It seems that in general the proportion of limb to trunk is less susceptible to plastic change than is overall body size.

These, and all the other studies that have been carried out along the same lines, confirm the existence of plasticity of migrants in response to the different environments to which they have journeyed and settled. It is of course not necessary to migrate to experience change of environment. Relevant variables can change in the one place: for example, food availability and quality, temperature and humidity, disease exposure, exercise and activity patterns. The food availability at different seasons of the yearly agricultural cycle with its accompanying change in labour demands is reflected in the cyclical adult physique differences, especially in weight, in many primitive cultivators, e.g. among the Azande (Culwick, 1953) the Massa (de Garine & Koppert, 1990) and in the Gambia (Fox, 1953) and in the weight-associated measures of skinfolds and circumferences. There is a similar cyclical pattern in child growth; in the Gambia, children grew very well in the dry season and poorly during the wet (and in the latter indeed growth in weight was often zero or slightly negative); trends in height were similar to those in weight (Thomson, 1977). Interpretation of the cyclical pattern in terms of diet and activity appears reasonable, although it is obviously more complex as shown by the small seasonal variations that occur in adequately nourished European children.

Experimental studies

Besides these empirical studies there are now numerous experimental investigations that show the nature, extent and mechanisms of plasticity. Some of these are nature's experiments, where observers exploit situations that have occurred without deliberate scientific planning and that illustrate a particular point, for example where some variables provide a contrast and others remain constant; others were intentionally designed and carried out; yet others arose where investigators monitored the effects of changes introduced for purposes other than experimentation (such as hygiene or nutrition).

Nature's experiments

At the beginning of the 1950s the adaptive significance of human physique variations attracted much attention. One question was how far the extreme physique of the Nilotics could be attributed to their way of life and other environmental factors. Comparison was therefore made of physique in three tribal samples of adult males who were living in virtually identical habitats but with minor differences in their ways of life (Roberts & Bainbridge, 1963). To provide a quantitative validation of the apparent differences in their way of life, haemoglobin level was measured and showed intertribal differences that corresponded to those in their cattle wealth (Roberts & Smith, 1957). As a genetic control, to exclude the possibility that physique differences were due to genetic differences among the groups, a genetic survey was carried out (Roberts, Ikin & Mourant, 1955); a range of gene frequencies showed that the tribes could be regarded as a homogeneous group. The results of an anthropometric survey (Roberts, 1977) were in the same direction as those in a somatotype study, showing slight differences between the tribes in the body measurements and physique. It was concluded that these slight physical differences between the tribes were to be attributed primarily to the effect on the phenotype of environmental differences, and particularly their ways of life and consequent dietary differences. Study of the dysplasia (Bainbridge & Roberts, 1966) suggested that the mechanism for the differences in adult physique was adjustment of growth rates at different stages to the environmental requirements.

A long-continuing examination of a wide range of physiological, physical and epidemiological variables in the Tokelau people was a particularly well-conceived study (Weesen *et al.*, 1992). Both the migrants to New Zealand and those islanders who remained at home on the atoll were examined several times over a long period. The way of life of both groups changed, and there was a considerable increase in weight in most age groups over the period of 19 years, but the weight gains were heaviest in the migrants. The body mass index adjusted for age in both migrants and non-migrants increased in both men and women, but remained consistently higher in the migrants in both sexes.

Similarly the study of the Samoans (Baker, Hannah & Baker, 1986) showed their response to changing conditions to be 'a massive weight gain', to which reduction in daily energy expenditure seemed to make an important contribution. Several other studies show that traditional peoples tend to gain weight as their way of life becomes modernized (Kasl & Berkman, 1983).

Bodily changes may be pronounced over quite short periods if the conditions are sufficiently extreme. Over a period as short as three weeks, soldiers undergoing strenuous physical activity during the course of their paratrooper training, although subjects in a good state of training initially, showed significant changes in body composition; body weight and body fat diminished, total body water, intracellular water and cell mass increased, while oxygen consumption also increased but remained constant relative to cell mass (Pascala *et al.*, 1955).

Attempts to colonize high-altitude regions show that 'the plasticity of the human response is impressive' (Shephard, 1991). Many of the reported differences of growth, adult stature and fat distribution attributed to inherited adaptations are on closer inspection more reasonably attributed to differences in socioeconomic status, dietary habits and living conditions. There is impressive evidence of physiological adaptation; for example, large static lung and heart volumes relative to body size, and the enlarged bone marrow and high haemoglobin level, for which the mechanism is well understood. In Native Americans these large highland chests are acquired through an accelerated development of the thorax relative to stature during childhood and particularly adolescence. However, these changes occur not only in those born at altitude, but also among those who move to high altitudes in early childhood (Mueller *et al.*, 1979). The large chest size of the high-altitude native is retained on moving to sea level.

Besides these observational studies of adults there are numerous studies monitoring supplementary feeding of children, part empirical, part experimental, for example in Baghdad (Mukhlis & Aikhashali, 1985), Haiti (Berggren, Hebert & Waternaux, 1985), Mexico (Richmond & Naranjo-

Banda, 1986), India (Devadas, 1986) and Colombia (Mora *et al.*, 1981). Children who received school meals or other supplementation to their diet were compared with a control group who did not; the children receiving the supplements grew significantly more in height and weight.

The effect of nutrition on growth in head size was well demonstrated by Stoch & Smythe (1963). They followed the growth of 21 very undernourished infants in South Africa from the age of one to five years. Better-nourished children of parents matched for head circumference with those of the malnourished sample were taken as controls. The growth in head circumference of the malnourished was less than the controls even at the time when the growth in both groups was nearing completion.

Designed experiments

As regards the soft tissue, responsible for so much of the appearance of the body, a classic study is what became known as the Minnesota starvation experiment under the direction of Ancel Keys, to which Lasker contributed. In 1944, 34 male volunteers were subjected to a famine diet for 24 weeks: a diet designed to simulate that of prison camps in Europe in World War II. Measurements were taken at the beginning and end of the experiment, and photographs for somatotyping were obtained. The measurements showed varying amounts of decrease, most in the circumferences with their large fat components and the least change in measurements of wrists, hands and feet where there is little soft tissue. Body weight decreased by 24%. The somatotype ratings showed that endomorphy diminished by an average of 47% and mesomorphy by 34%, and ectomorphy increased by 73%.

The mechanism by which the body changes in starvation are brought about is well established. In the first few days of starvation, the initial decrease of body mass is due mainly to fluid loss, associated with metabolic acidosis. The existing glycogen stores provide an initial reserve of carbohydrate. If the diet does not yield the essential minimum of carbohydrate to meet the requirements for brain and red cells of 80–90 g per day, glucose is formed by breakdown of protein (60–70 g per day), so that lean tissue contributes appreciably to the observed weight loss (57% of the weight loss in the Keys experiment (Keys *et al.*, 1950)). As the reserves of non-essential fat become exhausted, the protein is drawn on more and more, so that drain on the fat stores gives way to lean tissue depletion.

Another stimulus to changes in body composition and morphology is exercise. Regular exercise develops the appropriate muscles; the muscle cell proteins increase as a result of increased synthesis and decreased degradation so that the cells hypertrophy, there is increased muscle strength, and there is also increased concentration of aerobic enzymes affecting the rate of energy release. There are changes in the tendons, cardiovascular and ventilatory functioning, the nervous and digestive systems, and decreased deposits of accumulated fat. This response occurs at all ages, and was observed even in a study of nonagenarians.

Similarly, the mechanism by which changes in bone morphology are brought about is well known. Bones are remodelled as a result of the absorptive activity of osteoclasts working in harmony with the bone-forming activity of osteoblasts. In normal growth around the knee joints, as the shaft grows in length at the epiphysial plate, bone at the same time is removed from the surface of the metaphysis, so that the characteristic shape of the bone is retained. Superimposition of the outlines of the lower end of the femur in the sixteenth and twentieth year (Fig. 1.1) shows how much

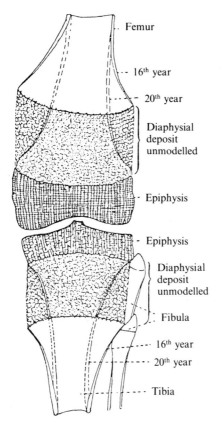

Fig. 1.1. Remodelling of bone around the human knee joint between the sixteenth and twentieth years of age. The bony tissue removed is indicated.

new bone is laid down and how much removed in the course of this remodelling during growth. Metrical evidence for this at later ages comes from studies of the radius in women aged 59–80 (Stini, Stein & Chen, 1992).

The actual process was illustrated many years ago by the use in the experimental situation of madder root ingestion: bone deposited during the feeding period is stained pink whereas that deposited in the madder-free period remains white, so it is possible to show exactly where bone deposition and bone absorption occurs during the process. In the pig, for example, Brash (1934) showed that the mandible increases in width as a whole by deposition of bone on its lateral surface and corresponding absorption from its inner surface. In the ascending ramus there is deposition at the posterior border and reabsorption from the anterior border, and in the horizontal ramus deposition of bone on the anterior surface of the symphysis gives growth in length. Remodelling occurs not only on the surface of bones during growth but also in the internal tissues, for the pattern of the bony trabeculae of cancellous tissue is continually being rearranged. This example illustrates the great plasticity of bony tissue. During development it may be deposited and removed again and again, so that in a mature structure nothing is left of the bony substance that was first laid down by the initial ossification. The process continues after maturity; indeed, its continuation into later life, with reduced coordination of deposition and resorption, may lead to the well-known increased risk of fracture in the elderly, and particularly in women.

The question is the extent to which this plasticity of bony structures is adaptive. There are clear instances where the changes brought about cannot be regarded as conferring an advantage, which is implicit in the concept of adaptation; the gross changes that come with severe under- or overnutrition may not be reversible when normal conditions return. However, there are numerous cases where these changes do seem to be advantageous. One example of apparent gross adaptation of bony structure is the curiosity of the little goat reported by Slijper (1942); this animal was born with phocomelia (undeveloped forelimbs) so that for locomotion it had to hop like a kangaroo, and with this mode of progression its pelvis became modified towards that of a biped (Fig. 1.2). The idea that the trabecular structure of cancellous bone was an adaptation to withstand tensile and compressive stresses dates from the previous century. Following the work of Wolff (1892), who pointed out that there is relationship between mechanical use and bone structure, it is generally thought that bone shape is related to its function and bone growth responds to local mechanical conditions. There is now much experimental evidence for the adaptive role of plasticity. In the classic series of *in vivo* biomechanical

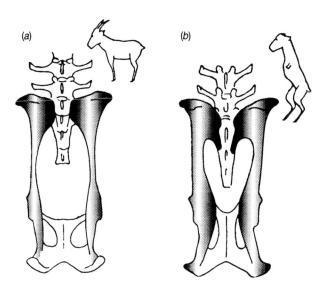

Fig. 1.2. Modification of pelvis in a bipedal goat.

investigations of primate mandibles by Hylander and his colleagues (1977, 1979a,b, 1984, 1985, 1987), strain gauges gave direct measurement of the stresses experienced in different regions of the bone. Radiographic studies (Mongini, Calderale & Barberi, 1979) in the human mandible demonstrated a consistent relationship between the direction of the experimentally induced principal strains with the direction of the bony trabeculae and the shape and orientation of the condyle. By computerized tomography, which allows the visualization of different sections of the bone, variations in cortical thickness can be measured, giving more precise estimates of bone strength than external dimensions (Moskowitch *et al.*, 1995). The mandible is remarkable for the density and thickness of its medial and lateral walls, and particularly where they coalesce at the base of the body. Yet despite its solidity, the mandible adapts itself to functional demands by remodelling to reflect the distribution of effective stresses, and so changes its mass, form and internal architecture.

Similar principles emerge from the studies of other bones. Bouvier (1985) examined tibial morphology in relation to stress during locomotion and showed that high anteroposterior bending stresses were associated with increased anteroposterior dimensions. In sheep calcanei, dorsal and plantar tracts of trabeculae corresponded to the principal compression and tension axes during the maximal weight-bearing portion of the stride. There is other evidence from pathology, for fractured bones, particularly in young individuals, not only repair but also remodel themselves so that the new shape is very similar to the original form. From his observations of healing of angulated fractures, Frost (1963) thought that bones actively drift in the direction of their concave surfaces in order to minimize bending, suggesting that axial compression is the predominant stress. Conversely, under diminished levels of bone strain, disuse atrophy is well known, and it appears from *in vivo* analyses that quite small reductions in function can profoundly affect cortical and trabecular bone volume in both younger and older individuals.

Conclusion

Plasticity is a word that has been applied and a concept that has been invoked in widely differing biological situations. In the sense of phenotypic response to environment, increasing plasticity in general characterizes the evolution of vertebrates, of mammals from reptiles, of primates among mammals, and of *Homo sapiens* from its primate ancestors, although plasticity in body form never reaches the extremes among animals which it does among many plants, and humans seem to be less plastic in body size than fish or mice. Most of the readily observable human traits are plastic to a greater or lesser extent.

From this review, plasticity is seen to apply not only to overall body form as earlier empirical studies suggested, but also to details of body composition, soft tissue distribution, and the structure of the bony skeleton. It occurs not only during the postnatal growing period but also during adulthood and senescence. It is, as Lasker noted, distinct from short-term physiological adaptation. Like other warm-blooded animals, the human individual adjusts physiologically to the immediate circumstances, experiences short-term acclimatization, and then readjusts when circumstances change. There is an extensive literature on human physiological responses.

Moreover, adaptive changes in one part of the body stimulate changes in others, as shown for example by the multiple responses to regular exercise. Such adaptation occurs as a result of performance of the function itself, and individuals vary in their capacity to adapt. This process of functional adaptation is so effective that it can bring about greater changes in one individual than can the most intensive genetic selection of hereditary variation over many generations (Pritchard, 1986).

An unexpected development from the concept of plasticity is the suggestion of an additional evolutionary mechanism, additional to natural selection. It is well known that some genes vary in their expression according to environmental factors, for example genes for local pigment synthesis in parts of the integument which vary in expression according to environmental temperature. Some genes also may be influenced according to the genetic constitution of the remainder of the genotype, so that an allele may have no noticeable effect in most genotypes, but will be expressed as a major effect if the background genotype is favourable. Waddington (e.g. 1967) showed that a temperature shock could induce an abnormal wing vein pattern in some *Drosophila* individuals. Breeding from those with this abnormal phenotype increased its frequency in subsequent generations. Eventually, individuals appeared with the defect without undergoing the temperature shock, and from these further breeding produced stocks in which the abnormal phenotype appeared at high frequency without heat shock treatment. These experiments showed that what appeared initially as a phenotypic feature caused by a clear environmental factor was being produced by the genome: in other words, there was a process of genetic assimilation. The explanation appears to be that the original population contained hidden genetic variation, which was revealed only under the abnormal condition of temperature shock. Under selection for its expression, the background genotype became reorganized so that a character that had originally been acquired became an inherited character that developed in the normal environment.

This then is Waddington's theory. It is based on the demonstration that variant alleles that are not expressed under normal conditions can be made manifest by exposure to abnormal stress, that selection of the responsive individuals increases the relative frequency of these alleles, and that individuals eventually appear who carry so many of the relevant modifying alleles that they manifest the resultant phenotype even before the stresses have been applied. Thus in natural selection, ancestors with the capacity for adaptation give rise to preadapted descendants. Waddington's theory of genetic assimilation is not yet well known, and still is not fully accepted. If confirmed, it would provide a genetic explanation for the very rapid capacity of species to develop morphologies and physiologies to meet the demands of their particular ways of life and their environments.

Plasticity is of fundamental relevance throughout biology. The concept, introduced into human biology I believe by Lasker, is one of the key links that hold the subject together, for it provides the rationale for widely disparate studies, which in their different ways are aimed at understanding biological variation in humans.

References

Bainbridge, D. & Roberts, D. F. (1966). Dysplasia in Nilotic physique. *Human Biology* **38**, 251–78.

Baker, P. T., Hannah, J. M. & Baker, T. S. (1986). *The Changing Samoans*. Oxford University Press.

Baxter, J. H. (1875). *Statistics, Medical and Anthropological, of over a Million Recruits*. Washington: Government Printing Office.

Berggren, G., Hebert, J. R. & Waternaux, C. A. (1985). Comparison of Haitian children in a nutrition intervention programme with children in the Haitian nation nutritional survey. *Bulletin of the WHO* **63**, 1141–50.

Boas, F. (1910). *Changes in Body Form of Descendants of Immigrants*. Senate document 208, 61st Congress, 2nd session. Washington, D.C.

Boas, F. (1912). *Changes in Body Form of the Descendants of Immigrants*. New York: Columbia University Press.

Bouvier, M. (1985). Application of *in vivo* bone strain measurement techniques in problems of skeletal adaptations. *Yearbook of Physical Anthropology* **28**, 237–48.

Bowditch, H. P. (1879). *The Growth of Children, a Supplementary Investigation*, pp. 35–62. Boston: State Board of Health of Massachusetts.

Brash, J. C. (1934). Some problems in the growth and developmental mechanisms of bone. *Edinburgh Medical Journal* **41**, 305–87.

Brezina, E. & Wastl, J. (1929). Anthropologische Konstitutions- und gewerbe-hygienisch Untersuchungen an Wiener Strasenbahnbediensteten. *Mitteilungen der Anthropologischen Gesellschaft in Wien* **59**, 19–38.

Culwick, G. M. (1953). Social factors affecting diet. *Transactions of the Sudan Philosophical Society* 1953, 1–19.

Devadas, R. P. (1986). Nutritional outcomes of a massive feeding programme in Tamil Nadu. *Proceedings of the Nutrition Society of India* **32**, 71–83.

Dobzhansky, T. (1957). *Evolution, Genetics and Man*. New York: John Wiley.

Ehrich, R. W. & Coon, C. S. (1947). Occipital flattening among the Dinarics. *American Journal of Physical Anthropology* **6**, 181–6.

Fishberg, M. (1905). Materials for the physical anthropology of the Eastern European Jew. *Annals of the New York Academy of Sciences* **16**, 155–297.

Fox, R. H. (1953). Energy expenditure of Africans engaged in various agricultural activities. PhD. thesis, University of London.

Frost, H. M. (1963). *Bone Remodelling Dynamics*. Springfield: C. C. Thomas.

de Garine, I. & Koppert, S. (1990). Social adaptation to season and uncertainty in food supply. In: *Diet and Disease*, ed. G. A. Harrison & J. C. Waterlow, pp. 240–89. Cambridge University Press.

Goldstein, M. S. (1943). *Demographic and Bodily Changes in Descendants of Mexican Immigrants*. Austin: University of Texas, Institute of Latin American Studies.

Gould, B. A. (1869). *Investigations in the Military and Anthropological Statistics of American Soldiers. Sanitary Memoirs of the War of the Rebellion. US Sanitary Commission*. New York: Hurd & Houghton.

Greulich, W. W. (1957). A comparison of the physical growth and development of American born and native Japanese children. *American Journal of Physical Anthropology* **15**, 480–515.

Harrison, G., Tanner, J. M., Pilbeam, D. R. & Baker, P. T. (1987). *Human Biology* (3rd edn). Oxford University Press.

Hulse, F. S. (1968). Migration and cultural selection in human genetics. *The Anthropologist*, special volume, pp. 1–21.

Huxley, J. (1942). *Evolution: the Modern Synthesis.* London: Allen & Unwin.

Hylander, W. L. (1977). *In vivo* bone strain in the mandible of *Galago crassicaudatus. American Journal of Physical Anthropology* **46**, 309–26.

Hylander, W. L. (1978). Incisal bite force direction in humans and the functional significance of mammalian mandibular translation. *American Journal of Physical Anthropology* **48**, 1–8.

Hylander, W. L. (1979a). Mandibular function in *Galago crassicaudatus* and *Macaca fascicularis*: An *in vivo* approach to stress analysis of the mandible. *Journal of Morphology* **159**, 253–96.

Hylander, W. L. (1979b). An experimental analysis of temporomandibular joint reaction force in macaques. *American Journal of Physical Anthropology* **51**, 433–56.

Hylander, W. L. (1984). Stress and strain in the mandibular symphysis of primates: A test of competing hypotheses. *American Journal of Physical Anthropology* **64**, 1–46.

Hylander, W. L. (1985). Mandibular function and biomechanical stress and scaling. *American Zoologist* **25**, 315–30.

Hylander, W. L., Johnson, K. R. & Crompton, A. W. (1987). Loading patterns and jaw movements during mastication in *Macaca fascicularis*: a bone-strain, electromyographic and cineradiographic analysis. *American Journal of Physical Anthropology* **72**, 287–314.

Ivanovski, A. (1925). Die anthropometrischen Veränderungen russischer Volker unter dem Einfluss der Hungersnot. *Archiv für Anthropologie* **20**, 1.

Kasl, S. V. & Berkman, L. (1983). Health consequences of the experience of migration. *Annual Review of Public Health* **4**, 69–90.

Keys, A., Brozek, J., Henschel, A., Mickelsen, O. & Taylor, H. J. (1950). *Biology of Human Starvation.* Minneapolis: University of Minnesota Press.

Lasker, G. W. (1952). Environmental growth factors and selective migration. *Human Biology* **24**, 262–89.

Lasker, G. W. (1954). The question of physical selection of Mexican migrants in the USA. *Human Biology* **26**, 52–8.

Lasker, G. W. (1961). *The Evolution of Man.* New York: Holt Rinehart & Winston.

Lasker, G. W. (1969). Human biological adaptability. *Science* **166**, 1480.

Lasker, G. W. (1976). *Physical Anthropology* (2nd edn). New York: Holt Rinehart & Winston.

Malina, R., Buschang, P. H., Aronson, W. L. & Selby, A. (1982). Childhood growth status of eventual migrants and sedentes in a rural Zapotec community. *Human Biology* **54**, 709–16.

Martin, W. J. (1949). *Physique of Young Adult Males.* London: HMSO.

Mascie-Taylor, C. G. N. (1984). The interaction between geographical and social mobility. In: *Migration and Mobility*, ed. A. J. Boyce, pp. 161–78. London: Taylor & Francis.

Mongini, F., Calderale, P. M. & Barberi, G. (1979). Relationship between structure and stress pattern in the human mandible. *Journal of Dental Research* **58**, 2234–37.

Mora, J. U., Herrera, M. G., Suescun, J., de Navarro, L. & Wagner, M. (1981). Effects of nutritional supplementation on physical growth of children at risk of malnutrition. *American Journal of Clinical Nutrition* **34**, 1885–92.

Moskowitch, M., Simkin, A., Gomori, J. M. & Smith, P. (1995). The relation between external dimensions of the human mandible and cortical bone morphology as determined with the aid of CT scans. *Human Evolution* (in press).

Mueller, W. H., Yen, F., Soto, P., Schull, V., Rothammer, F. & Schull, W. J. (1979). A multinational Andean genetic and health programme. *American Journal of Physical Anthropology* **51**, 183–96.

Mukhlis, G. M. & Aikhashali, M. J. (1985). Comparative study on the effects of school feeding programmes on the heights and weights of schoolboys. *Journal of the Faculty of Medicine of Baghdad* **27**, 89–94.

Ossenberg, N. S. (1970). The influence of artificial cranial deformation on discontinuous morphological traits. *American Journal of Physical Anthropology* **33**, 357–72.

Pascala, L. R., Frankel, T., Grossman, M. I., Freeman, S., Fuller, I. L., Bond, E. E., Ryan, R. & Bernstein, L. (1955). *Changes in Body Composition of Soldiers during Paratrooper Training*. Denver, Colorado: Medical Nutrition Laboratories. Report 156.

Pritchard, D. J. (1986). *Foundations of Developmental Genetics*. London: Taylor & Francis.

Richmond, J. A. Gonzalez & Naranjo-Banda, A. (1986). Effect of an integrated nutrition primary healthcare package on nutritional status of children. *Nutrition Research* **6**, 1275–80.

Roberts, D. F. (1977). Physique and environment in the northern Nilotes. *Mitteilungen der Anthropologischen Gesellschaft in Wien* **107**, 161–8.

Roberts, D. F. & Bainbridge, D., (1963). Nilotic physique. *American Journal of Physical Anthropology* **21**, 341–70.

Roberts, D. F., Ikin, E. W. & Mourant, A. E. (1955). Blood groups of the northern Nilotes. *Annals of Human Genetics* **20**, 135–54.

Roberts, D. F. & Smith, D. A. (1957). A haematological study of some Nilotic peoples of the southern Sudan. *Journal of Tropical Medicine and Hygiene* 1957, 45–52.

Shapiro, H. L. (1939). *Migration and Environment*. New York: Oxford University Press.

Shephard, R. J. (1991). *Body Composition in Biological Anthropology*. Cambridge University Press.

Slijper, E. J. (1942). Biological-anatomical investigations on the bipedal gait and upright posture in mammals. *Proceedings Koninklijke Nederlandsch Akademie van Wetenschapen, Amsterdam* **45**, 407–15.

Someswara Rao, K., Tasker, A. D. & Ramanathan, M. K. (1954). Nutritional haemoglobin surveys in children in Nilghiri district. *Indian Journal of Medical Research* **42**, 55.

Stini, W., Stein, P. & Chen, Z. (1992). Bone remodelling in old age: longitudinal monitoring in Arizona. *American Journal of Human Biology* **4**, 47–55.

Stoch, M. B. & Smythe, P. M. (1963). Does undernutrition during infancy inhibit brain growth and subsequent intellectual development? *Archives of Disease in Childhood* **38**, 546–52.

Tanner, J. M. (1981). *A History of the Study of Human Growth*. Cambridge University Press.

Thomson, A. M. (1977). The development of young children in a West African village. In: *Human ecology in the tropics*, ed. J. P. Garlick & R. W. J. Keay, pp. 113–26. London: Taylor & Francis.

Waddington, C. H. (1967). *Principles of Development and Differentiation.* New York: Macmillan; London: Collier Macmillan.

Walcher, G. (1905). Ueber die Entstehung von Brachy- und Dolichocephalie durch willkürliche Beinflussung des kindlichen Schadels. *Zentralblatt für Gynäkologie* **29**, 193–6.

Walcher, G. (1911). Weitere Erfahrung in der Willkürlichen Beeinflussung der Form des kindlichen Schädels. *Muenchener Medizinische Wochenschrift* **3**, 134–7.

Weesen, A. F., Hooper, A., Huntsman, J., Prior, I. A. M. & Salmond, C. E. (1992). *Migration and Health in a Small Community: the Tokelau Experience.* Oxford University Press.

Wolff, J. (1892). *Das Gesetz der Transformation der Knochen.* Berlin: Hirschwald.

2 Plasticity in early development

DORIAN J. PRITCHARD

He who sees things grow from the beginning will have the best view
of them.   *Aristotle*

Summary

Lasker (1969) identified three levels at which species undergo environmental
adaptation: selection of appropriate genotypes, phenotype modification
during ontogeny, and behavioural or physiological responses by fully
formed individuals. This chapter deals mainly with the second of these and
addresses recent evidence that both supports and challenges contemporary
beliefs.

The potential for expression of an entire genome is initially a property of
most cell types, but as ontogeny proceeds cells and tissues gain or lose
competence to respond to embryonic inducers, they acquire programmes
for future gene expression and are said to become determined. The
physiological properties, relative positions and volumes of tissues are
defined only roughly by the genome, their finer details resulting from
responses to mutual metabolic and physical interactions. Within cell
groups the plasticity of individual cells becomes overwhelmed by common
aspects of metabolism, the Leader Cell Hypothesis and the Community
Effect being attempts to elucidate the basis of this coordination.

As cytodifferentiation proceeds, specific patterns of methylation of
cytosine residues arise along the chromosomes, while acidic proteins and
modified histones associate with the DNA in the vicinity of those genes
concerned with specialized functions. The specialized properties of tissues
seem, however, to be controlled more by cytoplasmic than by nuclear
factors. The immune system behaves in an exceptional fashion, the system
as a whole finally acquiring an exquisite degree of plasticity although
individual cells suffer irremediable loss and rearrangement of gene segments
that ensures each clone can synthesize only one type of antibody molecule.

Definition of body pattern requires restriction of plasticity on a regional
or segmental basis and in this the orderly, localized action of protein
products of homeobox-containing genes plays a seminal role.

Experiments that address regeneration and transdifferentiation of ocular
tissues suggest that the differentiated phenotypes of embryonic cells are
maintained by physiological interactions with neighbouring cells of other

types. Transdifferentiation occurs when this interaction is disrupted and results in replacement of cell types that are absent. Determination of ocular tissues therefore occurs relatively late in embryonic life and follows overt cytodifferentiation. In a similar way, refinement of structure in the circulatory and skeletal systems follows adoption of function by those systems.

Within species that adopt predominantly the regulative, as distinct from the mosaic, developmental strategy, plasticity is seen as an essential factor in normal development. However, this same property makes developing organisms vulnerable to potent teratogens, as exemplified by the readjustment of homeobox gene expression patterns on exposure of embryos to retinoic acid.

Some deficiencies in neo-Darwinist evolutionary theory are resolved by the concept of genetic assimilation of the capacity to acclimatize.

Introduction

Apart from a very few exceptions, every cell in the body contains exactly the same complement of genes as every other cell. Epidermal cells, for example, are distinguished from nerve, kidney and all other cell types not by possession of a unique selection of genes, but by their expression of a unique selection. It is selective gene expression that confers the special properties that define the different cell types.

The switches that turn these genes on, regulate their expression up and down, or turn them off, operate for the most part from outside the cell. Some of those that control the epidermis are in the matrix that helps stick the cells to their neighbours. Some are in the basal lamina that underlies the proliferative layer. Some are transmitted from the underlying dermis. Some, as illumination, abrasion, pressure, hydration or dehydration, impinge on the skin from the outside world. Others arise as a result of the bending and stretching that accompanies the individual's physical activity. Phenotype, the sum of the observable and measurable properties of an organism, is the product of interaction between the genome it inherits from its ancestors and the environment in which it finds itself; localized aspects of phenotype are created by localized environments.

In very general terms, the potential for expression of any or all possible combinations of genes is initially a property of most cells, but as ontogeny proceeds, restrictions are placed on this potential. Tissues gain or lose 'competence' to respond to inducers; cells become 'determined', as the operation of sections of their genome is effectively closed down, while other sections are opened up for expression. Eventually, if all goes well, each cell type comes to perform the particular functions which in that individual's

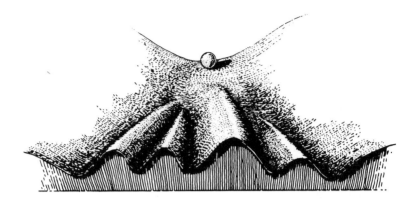

Fig. 2.1. The epigenetic landscape. The path of travel of the ball down the slope represents the course of development of a region of the embryo. Initially several possible pathways are open and any particular part of the embryo will follow one of these pathways. However, the further the ball travels down any one pathway the more difficult it becomes to move into an alternative one. (Reproduced from Waddington, 1957.)

ancestors assisted their survival, while other cell types harmoniously perform selected complementary functions.

As ontogeny proceeds, therefore, cells and tissues lose their capacity to respond, i.e. lose their plasticity, as they progress towards maturity. This global progression was visualized by Waddington (1957, 1966) as the 'epigenetic landscape' (see Fig. 2.1), an eroded slope down which a sphere, representing a cell or tissue type, inexorably rolls. Initially it has a choice of many destinations, but the further it progresses along any one route, the higher become the walls of its selected valley and the more difficult it becomes for its path to be changed.

Nevertheless, every type of tissue retains a limited capacity for regeneration or other modification which extends into adulthood, illustrating that some plasticity has in fact been retained. Such plasticity enables adult individuals to adjust to unusual conditions such as high altitude and brilliant sunlight, but it also renders them vulnerable to environmental hazards. In the early embryo, tissues are responsive to inductive influences that guide development in favourable directions, but their very responsiveness means that early development is disrupted with relative ease. Excesses of normal biochemicals like ethyl alcohol and vitamin A and its derivatives can cause malformation in human foetuses; synthetic chemicals such as the sedative and anti-emetic thalidomide can be even more harmful. Deficiencies of normal nutrients are also harmful. Deficiency of folic acid is implicated in causation of spina bifida and anencephaly. Susceptibility to teratogens and nutritional deficiency depends on the stage in ontogeny at which the embryo becomes exposed, but effects are not necessarily most severe in the earliest stages although errors that arise early are likely to have the widest consequences. For example, thalidomide causes malformation of the eyes and ears at 20–25 days, hypoplasia of the arms on days 26–30 and of the legs at 31–35 days (Insight Team, 1979). These periods presumably correspond to critical developmental stages of these particular organs.

Observations such as these show that in the development of the human embryo plasticity is important in normal differentiation and ontogeny, but in some situations may lead to pathology.

Modes of development

For a variety of ethical and practical reasons few experiments have been carried out on living human embryos, so most of our knowledge of prenatal plasticity in man is perforce derived by inference from other species. In the early days of experimental embryology, some of the most informative experiments involved nothing more than teasing embryos apart and observing the developmental capacity of the pieces. In some, like the limpets (*Patella* spp.), removal of one or two cells at the eight-cell stage caused irreparable damage. A quarter of a limpet, it seems, has the potential to develop only into that quarter it is predestined to become; the other three-quarters cannot make good its loss. In contrast, sponges can reorganize themselves to form one or several perfect little sponges even after passage of the whole organism through a fine sieve.

Such experiments led to the description of two contrasting modes of development, mosaic and regulative (see Hopper & Hart, 1980). In mosaic embryos specific developmental instructions are precisely located in the membranes or cytoplasm of the egg and after division these become associated specifically with individual nuclei. These nuclei then express only those characteristics defined by the instructions associated with them, without significant reference to the metabolic activities of neighbouring cells. In contrast, in regulative embryos, developmental instructions are less specific and operate over a large part of the body; tissues interact with one another, integrating their physiology and growth. In both modes of development many of the instructions are expressed as molecules with the property of binding to DNA and controlling transcription of pre-messenger RNA in that region of the chromosome. Some probably control RNA processing, the translation of RNA as polypeptide, assembly or modification of polypeptides to form enzymes and structural proteins, or the activities of enzymes.

Plasticity is therefore a property of less relevance so far as basic ontogeny is concerned to organisms that develop according to the mosaic principle, but organisms that begin their lives as regulative embryos are founded in plasticity, subject to the moulding forces that impinge on them from the outside world or arise as a result of the contortions of morphogenesis. Humans, it seems, are produced mainly by the regulative scheme, although the fates of cells may be sealed early on if they depend on the distribution of mitochondria, nutrients, or RNA or protein products of the maternal genome that were sequestered non-randomly in the ovum.

Cellular equivalence

In mammals each cell up to the eight-cell stage can form any part of the later embryo or adult. If one cell in a two-cell mouse embryo is destroyed, the other can develop into a perfectly normal animal. Monozygotic twins are created by early subdivision.

Two mouse embryos can be combined up to the eight-cell stage to form a single giant embryo, which, however, develops into a chimaeric mouse of normal size. Chimaeras can also be formed by injecting single donor cells into host embryos. In this way it is possible to assay the developmental potential of individual genetically tagged cells. Such experiments reveal that all cells are totipotent up to the eight-cell stage, but after that some developmental potency is lost. The reason for this is that the first three cleavage divisions leave all cells in topographically equivalent situations, their inner faces making contact with other blastomeres, their outer faces exposed and bearing whisker-like microvilli. Staining with fluorescent antibodies reveals protein constituents distributed asymmetrically in each cell. The fourth cleavage is orientated in a way that exploits this asymmetry, as inner and outer layers are formed, the inner lacking and the outer retaining the components associated with the microvilli (Maro et al., 1985).

Subdivision, fusion and microanatomy of early mammalian embryos all therefore demonstrate complete equivalence of cells up to the eight-cell stage.

In humans, monozygotic twinning may be considered a pathological event; if it occurs late or is markedly unequal it can create major developmental defects in some organs such as the heart. Mirror imaging of features such as handedness, finger and palm prints, and the position of hair whorls is a sign of late division.

Plasticity at the cellular level therefore starts to be lost at the 16-cell stage and arises as a result of non-random distribution of subcellular components in the parent cells. Embryos subdivided subsequently can develop normally provided critical components are represented in reasonable quantity, but subdivision that leaves a part embryo seriously deficient causes irreversible injury.

Acquisition and restriction of competence

As Georg Lichtenberg once wrote, it is astonishing that cats should have holes in their fur coats at exactly the right places for their eyes to see out. What is even more remarkable is that the lens of the eye is positioned at exactly the right place to cast a sharp image on the retina and that its optical properties change to maintain this relationship as the eye grows in size. The positioning of the different components of the eye did not arise, as many evolutionary geneticists believe, by selection of genes that precisely define those positions, but rather by selection of some genes that roughly define them and others that enable the tissues to control their relative growth and physiology by mutual interaction. This interaction is an aspect of *embryonic induction*, definable as communication between cells required for their cell-type-specific differentiation, morphogenesis, and maintenance. For cells to respond to induction they must have appropriate 'competence'. The eye has apparently evolved as, and develops as, an integrated system of mutually interacting tissues; it is this that ensures that all tissues are present in the correct places and functioning optimally (see Pritchard, 1986).

In the eye the most important interactions occur between a pair of lateral outgrowths of the embryonic brain, called the optic vesicles, and the overlying surface layer or ectoderm. The ectoderm at the point of contact develops into the lens, while the optic vesicle inverts at its tip to form the neural retina encased in pigment epithelium. The other parts of the eye assemble around the three major elements, neural retina, pigment epithelium, and lens.

Remarkably, eyes can form in other parts of the body if optic vesicles are implanted there. Experiments with frog embryos show that ectoderm that would normally later develop as skin on the flank and elsewhere, is competent at the early stages to develop instead into a lens. All it requires is an optic vesicle positioned close by, and the rest of the eye harmoniously forms as a result of their mutual interaction. Apparently competence to form the whole range of ectodermal derivatives, in appropriate species, teeth, hair, feathers, scales, even limbs, is initially widespread, but is lost as development proceeds. Reciprocal exchange of epidermis and dermis, the two layers in the skin, between feathered, unfeathered and scaled bird skin shows that characteristic differentiation of the epidermis is triggered by proximity to dermis from the appropriate region (Wessells, 1977).

Ectodermal competence is, however, not evenly distributed. Although lenses will differentiate in the ectoderm in many regions, the readiness with which this happens is enhanced at the paired sites where lenses normally arise. This localized enhancement of competence has been shown to be due to preinduction by other tissues, which ostensibly seem unrelated to the eye. For example, in salamanders one tissue that will later form the heart, and another destined to form the pharynx, are both in close proximity to the head ectoderm at an earlier stage and both contribute to enhancement of its lens-inductive competence.

Inductive interaction that initiates one course of differentiation seems to inhibit alternative courses, and the capacity for initiation of any course generally declines as the organism ages. The term 'determination' is used to describe the loss of plasticity that accompanies specialization and advancing age.

Inductive competence is therefore an aspect of embryonic plasticity that is developed and exploited as a normal feature of ontogeny, ensuring that neighbouring tissues adopt complementary forms and physiologies and ultimately that the whole body develops in harmony.

Nuclear equivalence and non-equivalence

Although the developmental potential of tissues becomes restricted as ontogeny progresses, that of the majority of the cell nuclei in those tissues may remain essentially unaltered. Evidence for this comes from experiments with frogs and toads in which nuclei were taken from differentiated body cells and inserted into enucleated eggs. When chromosome replication rates were coordinated with the cycle of the recipient cytoplasm, nuclei taken from differentiated cells of the gut, epidermis, and other tissues developed into perfectly healthy tadpoles (Gurdon, 1974).

When the nuclei of differentiated cells are examined by molecular biological techniques, however, subtle and complex changes are detectable. DNA can be cut by using highly specific enzymes, at specific short sequences of nucleotides. Some cut sequences that include adjacent cytosine(C) and guanine(G) molecules (in this context referred to as CpG) are represented also in reverse in the homologous strand of the double helix. The action of some of these enzymes (e.g. HpaII) is blocked if the cytosine carries an extra methyl (CH_3) group, whereas other enzymes (e.g. MspI) will cut the sequence irrespective of whether or not that cytosine is methylated. DNA was collected from differentiated tissues, divided into equal portions, and exposed to either one or the other enzyme. DNA fragments created by the enzyme action were then separated according to size by electrophoresis in a gel and, after Southern blotting, probed with DNA probes corresponding to the genes that code for tissue specific proteins. What such experiments revealed is that DNA in the region of the genes coding for specialized proteins, such as the globins and ovalbumin, is highly methylated in tissues where these proteins are not expressed, but unmethylated in those where they are expressed (see Razin & Riggs, 1980).

Methylation of cytosine is therefore a characteristic of non-expressed regions of the genome, whereas expression of specialized gene products is associated with non-methylation. Methylation is widespread in the DNA of early embryos and the differentiation of the cell types is accompanied by its selective demethylation in the different types of cell.

Demethylation of genes seems to be an aspect of determination, although it is probably a means by which differential patterns of gene expression become stabilized rather than a way in which such patterns are initially created.

In female mammals, although they have two X chromosomes, only one is active in any chosen cell; the inactive one is characterized by a high degree of methylation. The choice as to which X will be inactivated is made very early in development and all daughter cells derived from individual cells at that stage retain that choice. In contrast, methylation patterns of the non-sex chromosomes change as cytodifferentiation proceeds. It is suggested that this could be controlled by deactivation and reactivation of an enzyme, 'maintenance methylase', which copies the methylation pattern onto the new DNA strand every time the DNA is replicated at cell division. Incubation of cultured female cells with the drug 5-azacytosine destroys the pattern of DNA methylation and reactivates the inactive X. It also changes the patterns of expression of genes on the other chromosomes.

Other nuclear changes of a reversible nature include the association of specific acidic proteins and modified forms of the basic histone proteins with the DNA in the neighbourhood of those genes concerned with the specialized functions of that cell type.

The immune system is outstanding in its capacity to respond to exposure to molecules that are foreign to the body of that individual. This responsiveness is a property of the system as a whole; within individual B lymphocytes major changes that are irreversible take place and commit those specific lineages of cells to specific patterns of protein synthesis. This involves rearrangement and loss of DNA from the chromosomes. Specialization for specific antibody synthesis is therefore a one-way track so far as individual cells are concerned, but the system as a whole generally retains great lability.

Differentiation of lens cells and mammalian red blood cells involves loss of all nuclear materials, so that they are irredeemably differentiated once that pathway of differentiation has been followed. These examples are, however, exceptions to the general rule that cytodifferentiation occurs without major irreversible changes to the genome.

In general, then, plasticity of tissues with respect to potential expression of genes diminishes as development proceeds. In a few cell types this involves irreversible loss of DNA, but in the vertebrates a major factor of general importance is the establishment of patterns of (de)methylation of cytosine within or close to genes expressed tissue-specifically. Established methylation patterns are normally retained in progeny cells unless inductive influences intervene, but they may also be disrupted under some other circumstances, potentially causing loss of tissue-specific patterns of gene expression.

Regeneration and transdifferentiation

If determination were complete in mature tissues, wound healing could occur only by reorganization of undamaged tissue, or by multiplication of existing cell types, but in some species, notably the salamanders, spectacular regeneration can occur which involves interconversion of cell types, major restructuring and even replacement of whole limbs. Salamanders are unusual in that many members of the group show extreme paedomorphosis, sexual maturity being attained in a larval body. Some species mature physically only after exposure to thyroxine or iodine, which is absent from their native environment. It is the ratio of the two hormones thyroxine and prolactin that normally triggers metamorphosis in other amphibians. The regenerative ability of adult salamanders is therefore taken to indicate the capacity of embryos or juveniles of evolutionarily more advanced species such as our own and there is evidence that broadly this is probably true. Some of the most interesting observations are on the eye.

If the lens is removed from the eye of an adult salamander, within a few weeks a perfect new one regenerates at precisely the right position (Yamada, 1977). This generally forms from the pigment epithelium of the iris, a derivative of the optic vesicle. Conversion of pigment cells into lens involves loss of the differentiated characteristics of the pigment cells, including shedding of the melanosomes, a change in mitotic rate, synthesis of proteins characteristic of the lens, and ultimately loss of chromosomes and the major cytoplasmic structure, as occurs during normal lens differentiation. Apparently in the normal, intact adult salamander eye the differentiated state of the iris cells is maintained by the influential presence of the other tissues. When this is disrupted by removal of the lens, the metabolism of the iris pigment cells changes so as to adopt properties that return the eye as a whole to a state of physiological equilibrium.

If the neural retina is excised from a salamander eye, a new one will arise by conversion of cells of the retinal pigment epithelium. Again the eye apparently senses the loss of a tissue that helps maintain it in a state of internal equilibrium and takes steps to replace the missing component. Both lens and neural retina will regenerate like this if the original tissue is just displaced from its normal location. The eye therefore seems to sense, not just a general requirement for certain tissues to be represented, but also a requirement for their presence at specific locations and for them to be of a certain mass.

The same kind of cell conversion event occurs in eye tissues that are excised and removed from the influence of their normal neighbours. For example, if neural retina cells from chicken embryos are excised, disaggregated and cultured in isolation from other cells, some of them convert into pigment epithelium and some into lens cells, while others remain as neural cells. What was initially one tissue becomes a mixture of three. This differentiation is controlled by the biochemical pathway known as the tricarboxylic acid (TCA) cycle, concerned with the generation of energy (see Pritchard, 1981, 1986).

If retinal pigment epithelial cells from chicken embryos are cultured in an initially pure state, some convert into neural retina cells and others into lens cells. Again, what was initially one tissue becomes three.

This so called 'transdifferentiation' occurs in embryonic ocular cells of other species, including humans. In general it is inhibited by the cell type that would result from transdifferentiation and, whereas in salamanders this capacity continues throughout life, in most species it declines as the embryo matures.

Transdifferentiation seems to illustrate that in the growing eye the physical organization of the different tissues and the patterns of gene expression in their cells become established by physiological interactions between them (Fig. 2.2) and that those patterns normally become fixed by some other means as the animal matures. In salamanders the latter

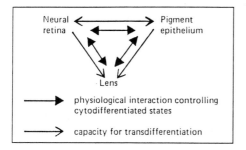

Fig. 2.2 The hypothetical tripartite tissue interaction system in vertebrate eyes. When all three tissue components are present the group is stable, owing to mutual interaction, but both neural retina and pigment epithelium will transdifferentiate into the other cell types if cultured in isolation.

condition is usually not achieved. It seems that differential patterns of gene expression therfore become established while the cells are in a state of metabolic plasticity and then become fixed by determination occurring at a later stage. This could involve extracellular changes. Mature tissues that have passed through the stage when they can transdifferentiate are more difficult to disaggregate, but they probably also have changes in their chromatin. Tissues that can transdifferentiate into lens are characterized by low levels of transcription of messenger RNA for lens crystallin proteins. Transdifferentiation into lens involves selective amplification of this activity; loss of the capacity to transdifferentiate accompanies its elimination.

Regeneration in the living eye and transdifferentiation in primary cultures of retinal cells illustrate several principles that relate to the harnessing of plasticity in the integration of tissue types. Firstly, they show that embryos and even their excised cells have an 'awareness' of being, or belonging to a physiologically balanced system of interacting components. Secondly, they show that patterns of gene expression in eye tissues are not inflexibly determined before expression, but are initially controlled physiologically before later becoming determined. Thirdly, they show that transdifferentiation occurs between cell types that are histologically very different, but anatomically close. This also implies that control of tissue-specific gene expression depends on the local tissue environment.

A full understanding of transdifferentiation and the restriction of transdifferentiative potential during maturation could pave the way to regenerative therapy of a type and degree that is at present quite beyond the bounds of possibility. Some of the most important tissue interactions involved in the creation of differential patterns of growth and gene expression in the embryonic eye seem to be retained into adulthood, but in human adults their influence is inhibited by the capacity of target tissues to respond. If these strictures on response in adult tissues could be relaxed, people with damaged retinas might be enabled to regenerate new ones, utilizing the directive forces already inherent within the eye, and so perhaps regain their sight.

Development through use

The idea that a bipedal, omnivorous ape naturally selected over millions of years for existence as a hunter–gatherer should have had conferred upon it by this experience the genetic endowment of a fighter pilot, a mathematician, or a poet seems absurd. Yet that is what has happened. The ingredient that makes it possible is the adaptability of individuals to their own personal circumstances, the behavioural equivalent of physiological acclimatization. The hunter–gatherer evolved an insatiable capacity to learn both mental approaches and physical skills; it is the adjustment of the nervous system in an individual, through challenge and training, that enables the success of our species in so many fields seemingly very far removed from those that challenged our ancestors. This capacity to master mental and bodily functions is probably enhanced in humans as a consequence of the evolutionary retention of characteristics associated with juveniles and the prolongation of childhood, with its consequent opportunities for training. Other species show comparable capacities to learn as juveniles, but proverbially you cannot teach new tricks to an old dog. Our capacity to train our minds and bodies towards new skills, even in old age, is therefore perhaps the neurological equivalent of the retention of embryonic regenerative properties in the bodies of salamanders.

Many aspects of our physique also are the products as well as the performers of function. It is customarily accepted that the plan of the body's blood system is defined in the genes, but this is probably only partly true as it has been shown in other species that quite abnormal, but perfectly functional, circulations result if selected blood vessels are ligated in the early embryo. The body, it seems, inherits not a strictly defined organizational plan, but a capacity to make the best of an ill-defined one, by internal adjustment to the forces and requirements that arise. This continues throughout life, but its effects are particularly dramatic at birth.

The circulation of the foetus at term has three notable specialized structures that are not present in a healthy adult. These are the placenta with its umbilical cord, a lung bypass called the ductus arteriosus, and a window between the two atria of the heart, called the foramen ovale. At birth the responsibility for supply of oxygen and elimination of carbon dioxide is transferred almost instantaneously from the placenta to the lungs. For this to occur the lungs must be in perfect working order before they are called upon to function, and the heart must arrange to redirect the flow of blood to the lungs instead of the placenta (Carlson, 1981). In the foetus, blood from all over the body is collected in a large vessel called the inferior vena cava and flows into the right atrium of the heart. Some then passes directly into the left atrium through the foramen ovale, then to the left ventricle in the normal way before expulsion into the dorsal aorta en route to the placenta. The remainder of the blood that enters the right atrium is squeezed in the normal way into the right ventricle, from where it is expelled through the pulmonary artery towards the lungs. However, instead of entering the circulation of the lungs it is shunted through the ductus arteriosus and thence into the dorsal aorta.

At birth resistance to blood flow in the lungs rapidly reduces, accompanied by closure of the ductus arteriosus, this being a response of the muscles surrounding it to increased oxygenation. Blood is therefore forced into the lungs, to emerge via the pulmonary vein and pass into the left atrium. The foramen ovale then becomes closed by a flap, which presses over it as a consequence of the new pressure differential on its two sides. This flap eventually grows in place and the lumen of the ductus arteriosus ultimately fills by overgrowth of connective tissue. Obliteration of the lumens of the umbilical vein and artery takes several weeks and occurs by tissue growth.

Dramatic changes therefore occur at birth triggered by changes in blood content. Once the new circuit is initiated, physical consequences ensure its reinforcement. Defects in heart function arise if the foramen ovale fails to close completely or if the ductus arteriosus remains open.

Observation of landmarks in ontogeny, like the emergence of milk teeth and permanent teeth and cessation of growth, indicate that, apart from the arrival of puberty, human development proceeds at something like half the pace of that of our closest relatives, the great apes (de Beer, 1962; Gould, 1977). Gestation is, however, only three weeks longer, so that human babies take 21 months from conception to reach the state of physical maturity of an average ape at birth. One effect is that the skeletal ossification that occurs during the latter part of pregnancy in apes takes place in man during the first year of postnatal life and there is a very big difference in the shape of the skeleton in the two species at the times when ossification takes place. In the unborn ape the spine is bowed and the legs tucked up in front, but at the comparable ontogenic stage the human baby can stretch out straight and kick its legs. The result is that our spines become hardened in a recurved shape instead of a bow and our pelvises solidify with the hip sockets directed downwards rather than forwards. These two features, the result of plasticity of the unossified skeleton, must have a great bearing on our ability to attain and maintain the characteristic bipedal stance which sets us apart physically from the other primates.

Refinement of physical structure therefore follows physical function just as determination of the patterns of gene expression in eye tissues follows their overt expression. Humans can reinforce or modify their physiques by adoption of postures or habitual physical movements. Likewise they can train their motor movements and the facility of their thoughts, exploiting the innate plasticity of what, by comparison with those of most other adult vertebrates, is essentially an immature nervous system.

Embryonic plasticity and evolution
The concept of plasticity helps resolve one of the oldest and most controversial issues in biology, concerning evolution. The big issue is to explain how animal species have evolved which are so superbly adapted to

their own specific environmental niches. The conventional neo-Darwinist explanation is that mutant genes arise spontaneously and these confer favourable phenotypic properties that are not shown by the parents or non-mutant brothers and sisters. Natural selection then ensures that the mutant individuals leave more descendants than the rest, until eventually virtually the whole population has the improved phenotype conferred by the new mutant gene. Conversely, if mutations arise that confer maladaptive phenotypes they are rapidly eliminated. Back in 1969, Lasker identified three levels at which adaptation occurs: selection of genotype, phenotype modification during ontogeny, and physiological and behavioural response. He pointed out that as one goes from differences between racial groups to differences between individuals within the human species, the chief emphasis shifts from the first level to the second and third. This is what one would expect from neo-Darwinist theory.

Historically the major opposing theory, espoused by the followers of Lamarck, included the idea that as animals strive to survive, their bodies become modified accordingly and such modifications are passed on to their descendants who then have a head start in coping with environmental hazards. This second theory conceives no mechanism by which acquired characters (apart from learned and taught behaviour) may be passed to offspring, but the first (which is generally accepted) also has major flaws. One of these is its failure to account for the acquisition of the observed number of favourable modifications of phenotype in the time available. The rate of acquisition of favourable mutations by genes such as those for the red blood pigment haemoglobin seems very much slower than the observed rate of change of gross phenotype observed in the fossil record. Secondly, conventional theory implies restriction of genetic variation in the population through inbreeding, which is perhaps incompatible with the high degree of genetic variation actually observed. Another striking but unexplained feature is the coexistence of great evolutionary conservatism of some traits alongside great lability in others. For example, the primitive vertebrate five-digit limb has evolved into a great variety of forms, ranging from the wings of bats and birds to the hooves of horses, the flippers of whales and the hands and feet of man. But despite this obvious capacity for change, in all these species the limbs have retained their basic pattern equivalent to humerous, radius and ulna, carpus, metacarpus and phalanges and their positions beside pectoral and pelvic girdles.

One of the biggest problems posed by animal evolution is the development of structures that appear to be adaptive responses to environmental stress long before appropriate stresses have been imposed on that individual. For example, the skin on the palms of our hands becomes thickened and reinforced with the protein keratin in response to abrasion and pressure. Such thickened skin occurs on the soles of our feet and becomes even more heavily keratinised if we do a lot of walking, but this thickened skin exists on the soles of babies' feet before they learn to walk and even before they are born. Ostriches have similarly keratinized regions on their legs and chest, which cushion them when they sit and which are present in the embryo before hatching. The calluses on the elbows of warthogs are similarly present before birth.

Waddington's explanation (Waddington, 1975) introduced the term 'genetic assimilation' and involved the concept that a genetically conferred capacity of some individuals to acclimatize favourably to new stresses is already present in non-stressed populations. When a new stress is imposed, those members with the genetically conferred capacity to adjust do so. This gives them a selective advantage and the capacity to respond appropriately is passed in the genes to their descendants.

The adaptive response is thus rather like a multifactorial threshold trait in which members of the population who are close to the threshold because they have an appropriate combination of genetic factors are pushed over that threshold by the additional effects of environmental stress. Experiments with fruit flies showed that breeding from individuals who before exposure to stress are close to, but below the threshold can eventually produce offspring who are above the threshold, even though unstressed (Fig. 2.3). According to the theory based on these observations, breeding from such individuals produces a race preadapted to specific environmental stress to which its members will probably subsequently be exposed. Responsibility for creation of the specific phenotypic feature has then been transferred from the environment to the genome, or the trait is said to have undergone 'genetic assimilation'.

In his review on human biological adaptability Lasker (1969) drew attention to the geographical distribution of skin colour and to individual variation in capacity to tan within populations of similar skin colour. Following Waddington's reasoning we can hypothesize that dark skin could evolve by selective breeding from those members of a lighter-skinned population that show the greatest capacity to tan, tanning being an acclimatic response that inhibits skin cancer and sunburn in brightly sunlit regions. Or conversely, less melanized skin could evolve from dark, pale skins conferring resistance to vitamin D deficiency in those areas of the world where exposure to sunlight is poor. The theory assumes that the genes responsible for a preadapted state, such as dark or pale skin, or thickened soles, are the very same as those involved in the adaptive response. Demonstration that they are not would require modification of the theory.

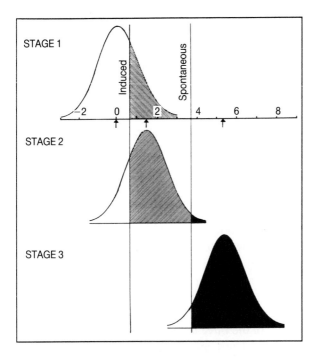

Fig. 2.3. Genetic assimilation of a polygenic threshold character. The curves represent the frequency distribution of a polygenic character along a scale of increasing expression, at three different stages in selection for that character. The vertical lines show the positions of two expression thresholds, induced and spontaneous. Individuals lying to the right of the spontaneous threshold express the character without environmental stimulus. Those that lie between the thresholds express it only on stimulation by a factor in the environment. Those to the left of both thresholds do not express it at all. The diagram illustrates the result of a real selection experiment for a feature of wing venation in *Drosphila* (after Falconer 1977; from data of Waddington).

Application of the concept of genetic assimilation to embryonic stages requires equation of acclimatization or physiological adaptation to stress by mature tissue with response to induction in an embryonic one. Capacity of the mature tissue to respond favourably to environmental stress now equates with competence of an embryonic one to respond to an inducer. With this insight, the evolution of new morphologies and new aspects of physiology which are laid down in the embryo can be conceived in terms of genetic assimilation operating at embryonic induction (Pritchard, 1986). The capacity to achieve a minor adaptive modification, for example in limb structure by appropriate use of the limb, is inherited as increased competence to respond to embryonic inducers that achieve the same result.

On this theory, in both the adult and the embryo evolutionary change in an adaptive direction is most easily achieved if the species inherits plasticity of phenotype. Indeed, if sufficient plasticity exists similar phenotypes can evolve from different starting points (convergent evolution), as it is the environment which defines what form an adaptive phenotype should take. When evolution occurs by this means the 'growing points' of an evolutionary trend are where maladaptation exists, where the greatest stress is placed on the bodies of a species' members. Phenotypes of individuals are therefore moulded by environmental stresses. Those individuals who respond most favourably pass on the genetic capacity for favourable response and eventually adaptive features of a similar type appear in descendants 'in anticipation of' that stress. Feathers probably first arose in the warm-blooded dinosaurs that gave rise to birds, as a response by the epidermis to external chilling, but in modern birds this external stimulus is replaced by an internal one arising from the underlying dermis. In modern birds the brood patches normally remain free of feathers, but feather growth can be stimulated there if dense material is placed beneath the epidermis to provide an internal stimulus.

Enhanced competence for lens induction in the head ectoderm is illuminating in that, of the three inducers known to operate on this tissue, pharyngeal endoderm, heart mesoderm and optic vesicle in that order (Jacobson, 1966), only the last bears a direct functional relationship to the lens in the mature eye. It is probable therefore that this last inducer in ontogenic terms was the first to evolve. We can thus visualize the evolutionary elaboration of phenotypic features as beginning in acclimatic responses at the interfaces between an organism and its environment, progressing through genetic assimilation to acquisition of similar features in the embryo and then progressing further, through genetic underpinning by appropriate preinduction and enhancement of competence, the latter operating also by genetic assimilation.

The theory as stated demands considerable genetic variation as it conceives genetic assimilation as deriving from the reassortment of alleles already largely present in the population. Genetic analysis of phenotypic traits shows that some that show genetic assimilation are polygenic, others monogenic. The latter can probably be accounted for in terms of incomplete penetrance and variable expressivity, the theory now being that selection is for alleles in the background genotype that enhance expression or increase penetrance. When assimilated these alleles assume responsibility for bringing the major genes to expression in place of the environmental factors that formerly had this role.

Incorporation of the concept of plasticity, with a genetic basis, into evolution theory therefore resolves major deficiencies of the neo-Darwinist model. Physiological adaptation, or acclimatization, has always been recognized by Darwinist philosophers, but they have been unwilling to accept it as the thin end of the wedge of evolutionary change, as Lamarck did. Instead they prefer to believe evolutionary change to be almost entirely the product of new mutation of nuclear genes. The theory of genetic assimilation of environmentally induced variation through selection for the capacity for appropriate response provides a respectable genetic mechanism that seems to resolve this ancient dispute.

Definition of body pattern

The most obvious and notable feature of an organism is what the early German embryologists called the Bauplan, which we know these days as the zootype or body pattern. Vertebrates in general have a single, segmented, major body axis, with a head at the fore end and two skeletal girdles, each providing the base for a pair of limbs. In humans the main axis distal to the pelvic girdle is truncated and carriage is upright. The pelvis and hind limbs are strengthened to carry all the body weight, while the fore limbs are slender, the digits elongated and the wrist freely rotatable. The means by which this pattern is defined are now becoming elucidated and it throws light on the role of genes and the modification of their effects by plasticity of response. The story is fascinating, but in order to understand it we need to know about early studies on a very, very distant relative, the darling of geneticists, the fruit fly, *Drosophila melanogaster*.

Despite its many obvious differences, the body pattern of *Drosophila* has notable similarities with our own. It has a single, major, segmented axis, with a head at the fore end; although it lacks skeletal girdles its limbs are, like ours, mounted in bilaterally symmetrical pairs. This differs very considerably from radially symmetrical, non-segmented species like jellyfish and sea urchins. Remarkably, the two segmented, bilaterally symmetrical patterns are founded on very similar genes which are present also in all other segmented species that have been examined. Astonishing as it may seem, the body patterns of all advanced forms of similar type seem to be derived from that of a very, very distant common ancestor, which existed before vertebrate evolution diverged from that of invertebrates. Despite separation by 500 million years of independent evolution, not only have the genes responsible for body pattern been conserved in their nucleotide sequences, but so have their linkage relations, in species as diverse as humans and flies.

Drosophila has some strange mutant variants in which organs are misplaced in the body. These are called homeotic mutants and include the *bithorax* mutant, which has two pairs of wings instead of one, and the *antennapaedia* mutant, which has legs where it should have antennae, as well as its normal legs. These organs normally develop from embryonic 'buds' called imaginal discs, which survive in the larvae in this form until metamorphosis. These acquire patterns of determination long before they grow and differentiate as adult structures; the fault in these mutants begins at the determination stage.

Determination of the imaginal discs occurs through exposure to morphogenic proteins that exist as gradients in the very early embryo; it is concentration-dependent. The gradients are established by the maternal system before the egg is laid and initially are represented as messenger RNA that codes for a gene-switching protein. This is translated in the embryo and the protein binds to specific sites in the DNA of nuclei, which at this stage are distributed uniformly in a monolayer just beneath the egg case. The DNA where binding occurs is arranged in tandem series of related structural genes and their respective on–off switch sequences or promoters. There are two series, one which defines the head, called the antennapaedia complex, the other which defines the thorax and abdoman, called the bithorax complex. An important feature of the original explanatory model (Lewis, 1978) was that the affinity of the gene-switching protein for a promoter site beside each gene increases sequentially along the series. The result is that different numbers of genes in the two complexes are switched on at different positions down the protein gradient, which extends down the length of the future body. The original model has now been extensively revised, but nevertheless determination of the different organs is still considered to occur as a result of expression of different combinations of genes defined initially by the gradient of maternally derived mRNA in the egg.

The genes of the antennapaedia and bithorax complexes themselves also code for proteins with gene-switching properties; some of these establish positive feedback loops controlling their own expression and also switch on other genes that code for organ-specific functions. Molecular analysis of the genes in this system reveals a sequence of 180 base pairs that is present in each, known as the homeobox. It is this sequence that allows the identification of comparable genes in man.

Not only is the homeobox present in many genes in our DNA, but groups of genes comparable to those of the antennapaedia and bithorax complexes and containing this sequence are also recognizable in their other nucleotides and their order side by side along our chromosomes (McGinnis & Krumlauf, 1992; Scott, 1992). In humans the whole system is represented four times as the HOX A (or 1), HOX B (or 2), HOX C (or 3) and HOX D

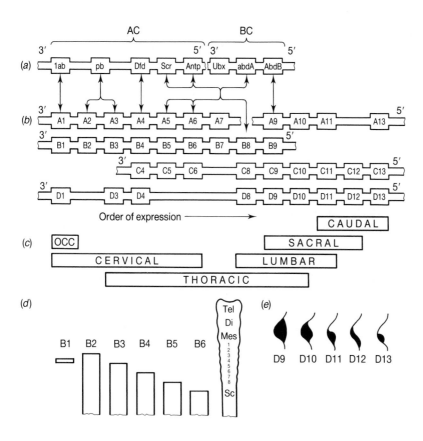

Fig. 2.4. Expression of homeobox-containing gene sequences in relation to body pattern. (*a*) Critical genes of the antennapedia (AC) and bithorax (BC) complexes in *Drosophila melanogaster*. The genes involved are *labial*, *proboscipedia*, *Deformed*, *Sex combs reduced*, *Antennapedia*, *Ultrabithorax*, *Abdominal A* and *Abdominal B*. (*b*) The paralogous genes in man of complexes HOX A, HOX B, HOX C and HOX D. (*c*) Expression of some genes of all four HOX complexes in the paraxial mesoderm in vertebrates. (*d*) Expression of HOX B genes in the embryonic hindbrain of the mouse. Tel, telencephalon; Di, diencephalon; Mes, mesencephalon; 1–8, rhombomeres 1–8; Sc, spinal cord. (*e*) Expression of HOX D genes in the chicken wing bud.

(or 4) groups. In our own species four genes are represented which have no equivalent in *Drosophila* and all except HOX B lack some of the *Drosophila* genes (see Fig. 2.4). Both human and *Drosophila* systems have probably evolved considerably since the common ancestor.

Utilization of the system in defining body pattern is well illustrated by expression of HOX B in the mouse hind brain (Hunt *et al.*, 1991; Holland,

Ingham & Krauss, 1992). Here are eight sequentially arranged sections called rhombomeres, the distinctions between which are believed to be defined by differential and orderly expression of genes in the HOX B complex. For example, if the individual genes are numbered sequentially, B2 is not expressed in rhombomeres 1 or 2, but is in number 3 and those posterior to it, B3 is expressed posteriorly from rhombomere 5, B4 from 7, B5 only posterior to rhombomere 8, and B6 only outside the hind brain. B1 is expressed only in rhombomere 4. Cranial nerves that emerge from these rhombomeres are similarly identified by HOX gene expression, as are the neural crest cells in this region which migrate out to form skeletal and other structures.

The HOX system also defines regionalization in the paraxial mesoderm (McGinnis & Krumlauf, 1992; Scott, 1992) although the pattern is less easily discerned as all four HOX complexes are involved. This gives rise to the vertebrae, ribs and scapulae and the musculature of the body wall, limbs, and dermis. The fore end of the body is specified by expression of genes at the lower-numbered (3') end of several HOX complexes and there is an orderly progression of expression towards the opposite (5') ends in tissue closer to the tail. Patterning in the tail in mice is specified by expression of HOX genes 11, 12 and 13, which have no homologues in *Drosophila*.

Definition of skeletal elements in the limbs involves the operation of several different systems, but here also the HOX system comes into play, producing overlapping regions of expression of HOX D genes (Duboule, 1992). This is centred on the zone of proliferating activity (ZPA) at the posterior margin of the limb bud, which defines anteroposterior patterning of the digits. In this case the genes correspond to the most posterior of the *Drosophila* bithorax complex plus those beyond it (D9–D13).

In *Drosophila* the genes of the antennapaedia and bithorax complexes are activated simultaneously, but to different degrees, relative to the concentration of morphogen in the gradient at that position. In vertebrates, however, activation of the series is sequential, as the primary organizer, Hensen's node, moves down the body axis at gastrulation. The HOX genes are therefore switched on in the order 1 to 13, the first to be expressed defining the head end, definition of the more posterior regions following as the node moves towards the tail and genes further along the HOX complexes come into expression. Sequential expression also occurs in the limb bud.

Body pattern in mice and fruit flies is therefore defined by utilization of very similar genes which have retained their relative positions for aeons, although their activation occurs by different means. In both vertebrates and insects orderly patterns of expression of genes become created that define subsequent development in the different segments or regions of the body. This involves selection of a restricted repertoire of options in each region so that within each region plasticity is reduced.

Biochemical modification of body pattern

Although strongly defined by genetic factors operating under control by developmental mechanisms many of which are also under genetic control, body pattern is nevertheless capable of drastic modification by teratogenic biochemicals. This can be seen as the negative face of plasticity. Of these teratogens one of the most potent so far known is retinoic acid, the acidic form of vitamin A.

The capacity of retinoic acid to change patterns of gene expression has been appreciated for some years (see Durston *et al.*, 1989), since a notable experiment involving exposure of teratocarcinoma cells. Teratocarcinomas are transplantable tumours of undifferentiated embryonic cells. They can be grown in culture as permanent cell lines and will proliferate indefinitely without differentiating, but they differentiate into a variety of apparently normal cell types if retinoic acid is added to the medium (Boncinelli *et al.*, 1991). In the cells retinoic acid associates with a receptor protein similar to the receptors for steroid and thyroid hormones. In this form it binds to specific DNA sequences and regulates transcription of RNA. When cultured teratocarcinoma cells were exposed to retinoic acid some of the HOX genes became activated. Activation occurred in the normal 3′ to 5′ order, but some at the 5′ end were not stimulated and although the activity of most genes was increased by retinoic acid, that of others was decreased.

Experiments with cultured mouse embryos showed that at seven days' gestation, retinoic acid causes an anterior shift in expression of the HOX genes (Kessel & Grüss, 1991). For example, the most posterior of the cervical and thoracic vertebrae were transformed into thoracic and lumbar types, respectively. When exposed at 8.5 days, however, the reverse effect was obtained. In the brains of mouse embryos exposed to retinoic acid, rhombomeres 2 and 3 were converted to 4 and 5; this also involved conversion of the fifth, trigeminal, nerve to the seventh or facial, and the neural crest cells supplying the face were similarly modified (Marshall *et al.*, 1992).

A classic experiment that demonstrates the source of a factor that normally specifies the digits involves excision of a block of tissue at the posterior margin of a chicken wing bud and implanting it into the anterior margin of another bud at a similar stage (Saunders, 1977). This causes mirror-image duplication in the resulting wing, an extra set of digits being produced in reverse order on the anterior side. Apparently some biochemical is emitted by the tissue at the posterior margin known as the zone of proliferating activity or ZPA, which guides the mesenchyme in the core of the wing bud in its differentiation as skeletal elements. The highest concentration defines the normally most posterior digit (number IV in the chicken wing), intermediate concentration defines digit number III, and the lowest concentration the most anterior (number II). The fascinating observation was that retinoic acid will substitute in this respect for the ZPA. When implanted at the anterior margin of a chicken wing bud, beads impregnated with retinoic acid also cause mirror-image duplication. Retinoic acid is also found to be present at equivalent concentrations in the normal limb bud and a good argument can be mounted that the natural morphogen emitted by the ZPA is in fact retinoic acid (Tickle, 1991).

The potential for teratogenesis by vitamin A derivatives in abnormal concentration is illustrated by the truly dramatic findings when a species of frog found in India, the marbled balloon frog (*Bufo andersonii*), was exposed to vitamin A palmitate added to the water of its aquarium after partial amputation of its tail at the tadpole stage (Mohanty-Hejmadi, Dutta & Mahapatra, 1992). This caused homeotic transformation of tail stump tissues, with growth of up to seven extra legs!

Bizarre though such experiments may seem, they provide a warning of the pathological consequences of exposing developmental systems to abnormal metabolites or even normal metabolites in abnormal concentrations. When its environmental milieu is drastically altered, the plasticity which under normal circumstances ensures correct development of a foetus can instead cause its downfall. In this context retinoic acid and related biochemicals must be considered particularly hazardous.

The Leader Cell Hypothesis and the Community Effect

Most biologists assume that all cells of similar age and appearance within a tissue are metabolically equivalent to one another, but we have no evidence that this is really true. Indeed when measures of any kind are made on individual cells a whole range of properties is usually found to underly the similar appearance of cells of the same histological type. A useful analogy is an army of uniformed men. Superficially they look and act more or less the same, but beneath his uniform each man has a unique personality with his own personal strengths and weaknesses. An army takes its lead from the directive of its commander and the orders of its officers and although this is an unconventional view, there is evidence that the properties of some tissues also may be governed by directives from a small proportion of leaders. For this to occur there must also be a large number of uncommitted 'follower cells' that do as they are instructed.

The first evidence for this theory came from observations on transdifferentiation of chicken embryo neural retina into pigment epithelium. When neural retina cells transdifferentiate in culture they go through several morphologically distinguishable stages. Melanin pigment is first detectable in a few scattered foci of 'pigment leader cells' within the colourless sheets of 'potential pigment cells' and from these it spreads outwards until eventually each sheet is evenly pigmented. The pigment leader cells usually die, but this has no effect on the outward spread of pigmentation. The number of pigmented foci increases if the activity of the tricarboxylic acid (TCA) cycle is slightly depressed, but TCA cycle depression has no effect on the spread of pigment (Pritchard, 1981). This suggests that the pigment leader cells are uniquely responsive to TCA cycle depression and that, having themselves modified their physiology accordingly, they then influence their immediate neighbours to follow suit, but they do this by a modification of the events by which they themselves reached that state. Leader cells are detectable also in other systems. For example, the pulsing of the cells of the embryonic heart is coordinated by 'pacemaker' cells that dominate the rate of pulsing of their neighbours.

According to the Leader Cell Hypothesis (Pritchard, 1986) an undifferentiated tissue such as the ectoderm in an early frog embryo may contain a range of leaders of diverse potential scattered through a mass of pluripotent follower cells, the range reflecting the inductive competence of the tissue. On exposure to an optic vesicle, lens leaders respond by changing their physiology towards that of lens cells and then influence their neighbours to do likewise. In regions where a different inducer is active a different set of leaders responds and they lead the followers along a separate differentiative path. If the followers remain undetermined for a considerable time after adopting the differentiated physiology they could be capable of transdifferentiation to another cell type, but normally only if separated from their original leaders and exposed to conditions that mobilize leaders of an alternative predisposition. Tissue disaggregation and reaggregation are usually recognized as normal prerequisites for transdifferentiation to take place.

The extension of phenotype from a cell to its neighbours may occur in two ways, either by growth of the leaders' extracellular matrix around or beneath adjacent cells, or by direct transfer of small effector molecules through gap junctions (Pritchard, 1992). In either case it is suggested that this process cuts short the metabolic events that initiated differentiation of the leaders.

Loss of inductive competence could involve maturation of leader or follower cells beyond their initially responsive states; determination could involve fixation by means such as cytosine methylation and synthesis of stabilizing extracellular matrices.

An effect that is possibly related to the secondary spread of pigmentation in neural retina cultures is what is known as 'the Community Effect', in which cytodifferentiation of a particular type occurs only in groups of contiguous cells that exceed a certain minimum number. Transdifferentiation into lens cells is stimulated by sowing neural retina cultures at high density, by aggregration, or by folding the cell sheet, but the clearest example of a community effect is provided by the work of Gurdon and colleagues (1993) on development of muscle cells in amphibians.

When muscle progenitor cells from early embryos of *Xenopus* are cultured as the filling in an ectoderm sandwich, myogenic and muscle-specific genes are activated within them, but only if the cells are in large aggregates. The cut-off point is around 100 cells. This therefore seems to be a social effect, which ensures that cells do not differentiate in isolation; interestingly, it occurs without mitosis or gap-junctional communication between its members. It seems to be purely the result of mutual interaction between large numbers of cells of similar competence, but how it works and whether there is an upper limit on such communities is as yet unknown. Possibly transdifferentiation would be the outcome if upper limits were exceeded.

The Leader Cell and Community Effect concepts are both attempts to elucidate the observation that adoption of a differentiated phenotype occurs in groups of contiguous cells. The observations on embryonic muscle relate to cell numbers, but not to the timing of changes in community members, whereas the Leader Cell Hypothesis deals with timing rather than cell numbers. The pigment leader cells in transdifferentiating neural retina cultures achieve the pigmented state in groups of 3–4 swollen cells or around 20 'potential pigment cells' of normal size; their differentiation could be the result of mutual interaction within those small communities. They are observed only within extensive sheets of potential pigment cells, but as yet we have no evidence as to whether or not these sheets have a minimum size. Both theories imply plasticity in the undifferentiated cells and coordinated loss of plasticity as differentiation ensues.

Lasker's (1969) use of the term 'plasticity' was with reference to the modification of an individual during its growth and development, as distinct from those aspects of adaptation that are genetically entrenched in the population by repeated natural selection and those dependent on a capacity to acclimatize over a short period. I have considered the term to embrace the latter concept together with his usage and I have suggested that all three aspects are involved in the evolutionary acquisition of superior fitness by a population or species.

Conclusion

Plasticity of physiology and form obtain throughout life and that which remains in older adults is only a much depleted remnant of embryonic plasticity. Of the two developmental modes, mosaic and regulative, human ontogeny proceeds mainly in accordance with the latter, in which plasticity is a central feature. Selective expression of the genes in the tissues depends to a large extent on the physiological stimuli to which they are exposed; these vary from tissue to tissue. Competence, the enhanced capacity to respond to inductive stimuli by changes in gene expression, is an elaboration of the basic responsiveness of embryonic tissues to changed conditions. Competence and determination, the associated restriction in responsiveness, probably involve methylation and demethylation of cytosine residues around critical genes.

Normal development requires the influence of some of the physiological concomitants of function, and the transdifferentiative changes that occur when some embryonic cells are separated from their normal neighbours implies that spatial organization within the body depends on mutual physiological interactions between tissues in proximity.

Plasticity is therefore important in development both in the response of tissues to influences external to them and in the generation of influences that create responses in other tissues. Through genetic assimilation it also provides a mechanism for preadaptation, enabling adaptive evolution to occur at speeds much greater than could be expected from mutation alone.

At some stage in ontogeny the extreme plasticity of embryonic systems must be subjugated in the interests of stable cytodifferentiation. In this, rigid definition of body pattern by highly conserved sets of HOX genes expressed in strictly controlled fashion, and leader cell and cell community effects, are probably important factors.

One of the widespread myths of modern biology is that the DNA of an organism contains all the instructions required to make that organism. This is incorrect for three reasons. The first is that vital details such as the structure of the plasma membranes, the mitochondria and even glycogen are not defined by genes, but grow more like crystals, taking their form from previous structure. The second is that the information encoded in the DNA is only useful if read in a specific physiological context. The gene that codes for the enzyme that converts the amino acid tyrosine into DOPA and then DOPAquinone, as a precursor for melanin pigment, is ineffective if tyrosine is absent owing to dietary deficiency, or the pH is wrong, or the cell is not at an appropriate temperature. The third reason is that when considered at the level of gross phenotype the message provided by the DNA is only a guide. The DNA codes for a highly complex, but adaptable, plastic system, which responds in many ways to the exigencies of survival. If the collection of structures and aspects of physiology defined by the genes is likened to an orchestra supplied only with the score to a major theme, the acclimatic changes that occur in the mature individual and the inductive responses of embryonic tissues, both aspects of their basic plasticity, can be thought of as the intricate and never-ending variations that provide much of the mystery and magic of the majestic symphony of life.

References

de Beer, G. (1962). *Embryos and Ancestors.* Oxford: Clarendon.

Boncinelli, E., Simeone, A., Acampora, D. & Maniolo, F. (1991). Gene activation by retinoic acid. *Trends in Genetics* **7**, 329–34.

Carlson, B. M. (1981). *Patten's Foundations of Embryology.* New Delhi: Tata McGraw-Hill Publishing Co.

Duboule, D. (1992). The vertebrate limb: a model system to study the *Hox*/HOM gene network during development and evolution. *Bio Essays* **14**, (6), 375–84.

Durston, A. J., Timmermans, J. P. M., Hage, W. J., Hendricks, H. F. J., de Vries, N. J., Heideveld, M. & Nieuwkoop, P. D. (1989). Retinoic acid causes an anteroposterior transformation in the developing central nervous system. *Nature* **340**, 140–4.

Falconer, D. S. (1977). *Introduction to Quantitative Genetics.* London: Longman.

Gould, S. J. (1977). *Ontogeny and Phylogeny.* Cambridge, Massachusetts: Belknap.

Gurdon, J. B. (1974). *The Control of Gene Expression in Animal Development.* Harvard University Press/Oxford University Press.

Gurdon, J. B., Tiller, E., Roberts, J. & Kato, K. (1993). A community effect in muscle development. *Current Biology* **3**, 1–11.

Holland, P., Ingham, P. & Krauss, S. (1992). Mice and flies head to head. *Nature* **358**, 627.

Hopper, A. F. & Hart, N. H. (1980). *Foundations of Animal Development.* Oxford University Press.

Hunt, P., Gulisano, M., Cook, M., Sham, M.-H., Faiella, A., Wilkinson, D., Boncinelli, E. & Krumlauf, R. (1991). A distinct *Hox* code for the branchial region of the vertebrate head. *Nature* **353**, 861–4.

Insight Team of the Sunday Times (1979). *Suffer the Children: The Story of Thalidomide.* London: Andre Deutsch.

Jacobson, A. G. (1966). Inductive processes in embryonic development. *Science* **152**, 25–34.

Kessel, M. & Grüss, P. (1991). Homeotic transformations of murine vertebrae and concomitant alteration of *Hox* codes induced by retinoic acid. *Cell* **67**, 89–104.

Lasker, G. W. (1969). Human biological adaptability. *Science* **166**, 1480–6.

Lewis, E. B. (1978). A gene complex controlling segmentation in *Drosophila. Nature* **276**, 565–70.

Maro, B., Johnson, M. H., Pickering, S. J. & Louvard, D. (1985). Changes in the distribution of membranous organelles during mouse early development. *Journal of Embryology and Experimental Morphology* **90**, 287–309.

Marshall, H., Nouchev, S., Sham, M. H., Muchamore, I., Lumsden, A. & Krumlauf, R. (1992). Retinoic acid alters hindbrain *Hox* code and induces transformation of rhombomeres 2/3 into a 4/5 identity. *Nature* **360**, 737–41.

McGinnis, W. & Krumlauf, R. (1992). Homeobox genes and axial patterning. *Cell* **68**, 283–302.

Mohanty-Hejmadi, P., Dutta, S. K. & Mahapatra, P. (1992). Limbs generated at site of tail amputation in marbled balloon frog after vitamin A treatment. *Nature* **355**, 352–3.

Pritchard, D. J. (1981). Transdifferentiation of chicken embryo neural retina into pigment epithelium: indications of its biochemical basis. *Journal of Embryology and Experimental Morphology* **62**, 47–62.

Pritchard, D. J. (1986). *Foundations of Developmental Genetics.* London: Taylor & Francis.

Pritchard, D. J. (1992). Unstable cytodifferentiation. In: *Fundamentals of Medical Cell Biology*, vol. 7, *Developmental Biology*, ed. E. E. Bittar, pp. 57–101. Greenwich, Connecticut: J. A. I. Press.

Razin, A. & Riggs, A. D. (1980) DNA methylation and gene function. *Science* **210**, 604–10.

Saunders, J. W. Jr (1977). The experimental analysis of chick limb bud development. In: *Vertebrate Limb and Somite Morphogenesis*, ed. D. A. Ede, J. R. Hinchliffe & M. Balls, pp. 1–24. Cambridge University Press.

Scott, M. P. (1992). Vertebrate homeobox gene nomenclature. *Cell* **71**, 551–3.

Tickle, C. (1991). Retinoic acid and the chick limb bud development. *Development* (suppl), **1**, 113–21.

Waddington, C. H. (1957). *The Strategy of the Genes.* London: George Allen and Unwin.

Waddington, C. H. (1966). *Principles of Development and Differentiation.* New York: Macmillan; London: Collier-Macmillan.

Waddington, C. H. (1975). *The Evolution of an Evolutionist.* Edinburgh University Press.

Wessells, N. K. (1977). *Tissue Interactions and Development.* Menlo Park, California: W. A. Benjamin.

Yamada, T. (1977). *Control Mechanisms in Cell-Type Conversion in Newt Lens Regeneration.* Basle: Karger.

3 Plasticity in the growth of Mayan refugee children living in the United States

BARRY BOGIN

Summary

The present-day Maya of Guatemala are characterized ethnically by many biocultural features. One is short stature. It is sometimes asserted that Mayans are 'genetically short' owing to generations of adaptation to an environment of poor health and nutrition. Alternatively, populations like the Maya are described as 'small but healthy'. These notions imply that no amount of intervention or economic development is needed for the Maya. Recent migration of Mayan refugees to the United States affords the opportunity to study the consequences of life in a new environment on the growth of Mayan children. The results of this research show that the notions of 'genetic shortness' and 'small but healthy' for Mayans are incorrect, and that an improved environment for health and nutrition leads to increased growth.

The children of this study live in Indiantown, Florida, and Los Angeles, California. Mayan children between 4 and 12 years old ($n = 240$) were measured for height, weight, fatness, and muscularity. Overall, compared with reference data for the United States, the Mayan children are, on average, healthy and well-nourished. They are taller, heavier, and carry more fat and muscle mass than Mayan children living in a village in Guatemala. However, they are shorter, on average, than children of Black, Mexican-American, and White ethnicity living in Indiantown.

These findings are typical of migrant adaptation to new environments as described by Boas in 1912 and elaborated into a formal model of human biological plasticity by Lasker in 1969. Lasker's model does not clearly differentiate between plasticity that is the result of a change in the environment and plasticity that leads to a new adaptive state. The variation in Mayan stature observed in living populations and in the archaeological record is interpreted as a response to environmental change, not biological adaptation. Finally, it is concluded that the Mayan refugees to the United States are in the process of a secular trend in height growth. Within a few generations, United States-born children of Mayan descent, if allowed equal access to health, economic, and social opportunities, should have an average stature similar to that of the general North American population.

Introduction

In 1969 Gabriel Lasker defined three types of human biological adaptations. The first and second are: '... those genetically entrenched in the population by repeated natural selection and those dependent on a capacity to acclimatize in the short run...' (1969, p. 1484). Lasker characterized the third type of adaptation as '... modification of an individual during his growth and development... the process is essentially irreversible after adulthood, ... and may be separately designated as plasticity' (1969). Although the term 'plasticity' had been used earlier to describe the malleability of human morphology during growth and development, it was Lasker's definition, as juxtaposed with those for genetic and acclimatized adaptations, that operationalized the concept of plasticity in the human biological sciences.

Lasker also showed that the mechanism for plasticity in human biology is to be found in those processes that regulate amount of growth and rate of development of the body as a whole, body segments, organs, tissues, or even populations of cells. In subsequent publications Lasker tried to show that the concept of plasticity could also be applied to human demography (Lasker, 1984). However, scant attention was paid to the concept of plasticity (see Chapter 11). Two reasons may be given for this. The first is that it was not clear whether the process of developmental plasticity results only in a change in the adult phenotype, or in a biologically significant adaptation. Chronic undernutrition, for example, usually reduces adult stature. Is this merely due to a lack of raw material for growth, or is this an adaptation in the strict biological sense in that the energy diverted from growth is saved for reproduction in adulthood? The second reason why the plasticity concept was not immediately popular is that in 1969 relatively little was known about the complexities of the endocrine system and its interactions with genes and environments that regulate growth control in the individual. Nor was much known about the mechanisms of demographic regulation. In the decades since 1969 much has been learned of these mechanisms of growth control and population dynamics. Now it is possible to evaluate the effectiveness of the mechanisms producing biological plasticity as well as the usefulness of the plasticity concept.

Some history of the plasticity concept for human growth

Perhaps the earliest American usage of 'plasticity' by a biological anthropologist was by Franz Boas. In 1910 he submitted an abstract of his work on *Changes in the Bodily Form of Descendants of Immigrants* to the United States Congress, which had commissioned the work. Boas measured thousands of immigrants from southern and eastern Europe and their children born in the United States. Boas found that the children of immigrants to the United States were taller than and differently shaped from their parents and the non-migrating populations from which their parents came. He stated that neither natural selection nor heterosis could adequately account for these changes. Rather, modifications in the process of growth and development as a response to environmental change were responsible. Accordingly, Boas concluded that 'we must speak of plasticity (as opposed to permanence) of types' (1910, p. 53). The term 'types' here means genetically fixed sizes and shapes of the human body or parts of the body.

Migration as the paradigm for plasticity studies

In a later discussion of his report to the U.S. Congress Boas stated that, '. . . we do not know the causes of the observed changes . . .' (1940, p. 71). To explore possible causes Boas (1912, 1922) cites studies by Livi (1896) and Ammon (1899) that treated rural-to-urban migrants. With this review, and his own work on immigrants to the United States, Boas established the study of human biological response to rural-to-urban migration as the paradigm for all subsequent work on human biological plasticity. Thus, it is worthwhile to summarize briefly what Boas discovered. Livi found that the children of rural-to-urban migrants in Italy were taller than rural sedentes. He believed that the reason for this was heterosis, the marriage of migrants from different rural regions leading to 'genetic vitality' in their offspring. Ammon also found the children of migrants to be taller than rural sedentes, but he argued for the action of natural selection to explain this. Although Ammon is not clear as to the agent of natural selection, perhaps he meant that in the rigors of the urban environment only the 'fittest' (the tallest?) would survive.

Livi's and Ammon's speculations that heterosis or natural selection were at work stem directly from their erroneous belief that human types were genetically fixed and that types would not change when exposed to different environments. That belief was shattered by the publication of Boas' studies of rural-to-urban migrants in the United States. Boas was vigorously attacked for his conclusion that human morphology was plastic during growth and development. In a series of papers he whittled away at his attackers (reprinted in Boas, 1940). His evidence against the fixity of types was that: (1) the physical differences between parents and children appear early in the life of the child and persist until adulthood; (2) the longer the childhood exposure to the American urban environment, the greater the physical difference; (3) children from large families (i.e. poorer families) are shorter than children from smaller families of the same 'racial type'; and (4) differences between parents and children are greater when both were foreign-born. In this case, the significance of the differences lies in the fact that only the children would have been exposed to the new environment during the developmental years. In the event that the child was American-born, the parents may also have spent some of their growth years in the United States as well.

The now classic studies of Shapiro (1939) on the growth of Japanese children in Japan and Hawaii, and of Goldstein (1943), Lasker (1952), and Lasker & Evans (1961) on the growth of Mexicans in Mexico and the United States confirmed the nature of human developmental plasticity during growth. Today, human developmental plasticity is taken for granted, but it was the European rural to American urban migration data of Boas that established the validity of this phenomenon.

The present study

The new research on human plasticity continues in the intellectual and methodological tradition started by Boas and refined by Lasker. This new work is a study of plasticity in the growth of ethnic Mayans who are the children of immigrants from Guatemala to the United States. The present-day Maya of Guatemala are the cultural descendants of the ancient Mayan civilization that occupied southern Mexico and Central America before the arrival of the Spanish in the early 1500s. Mayan culture is characterized by traditional clothing styles (often hand-woven and embroidered), social behavior (including family, marriage, and religion), and language (there are 22 major Mayan languages in Guatemala) that pre-date the Colonial invasion. Mayans also believe that they have *sangre de la raza*, meaning blood of the ancient Mayans, running in their veins. These cultural traits identify Mayans as one of the two dominant ethnic groups in Guatemala. Members of the other ethnic group are called *ladinos*. In contrast to the Maya, *ladinos* wear modern-style European or North American clothing, practice social behavior derived from the Spanish *conquistadores*, and speak Spanish. They also claim to be descendants of Spanish or other European peoples, i.e. without *sangre de la raza*. As of 1980, between 3.75 and 4 million of Guatemala's 8 million people were ethnically Mayan.

In addition to cultural traits, physical features have also been used to define Mayans. One such feature is stature: Mayans are often considered to be a short-statured people. Although it is true that the present-day Maya are, on average, shorter than other groups of people, the reasons for these height differences are not known. It has been proposed that Mayans are 'genetically short', that is, each generation of children inherits its small stature from the preceding generation. During fieldwork in Guatemala, the present author often heard this explanation from university-educated *ladinos* and occasionally even from Mayan men and women. Thus, 'genetics' is the popular explanation for Mayan short stature, a view also expressed by some foreign scientists and writers. One writer of popular science calls the Mayans a 'pygmy' people of the Americas (Diamond, 1992). Diamond, a human physiologist and ornithologist, writes that, '...several unrelated peoples ... evolved small size independently' (1992, p. 73). By 'evolved' Diamond means acquired a genetic characteristic produced by natural selection. Among the 'several unrelated peoples' are central African Pygmies, '... Bushman of southern Africa ... the Maya and other small-sized American Indians who are arbitrarily classified as Pygmies because their adult men measure under 4 feet 11 inches' (1992).

These views are too simplistic and, in part, factually incorrect. Growth surveys show that Mayans are not 'pygmies'. By definition, pygmies are biological populations in which men average less than 150 cm tall as adults. Mayan men average 156 cm to 169 cm tall in the few studies that have been published (Shattuck & Benedict, 1931; Steggerda & Benedict, 1932; Crile & Quiring, 1939; Goff, 1948; Méndez & Behrhorst, 1963; Bogin, Wall & MacVean, 1992). The reports of Mayan stature from the 1930s and 1940s state mean heights to be between 156 and 159 cm. However, the measurements were taken from men of uncertain age. Many subjects were listed as under 20 years of age. A recent longitudinal analysis of Mayan growth estimates that height growth continues into the early twenties for Mayan men (Bogin *et al.* 1992); thus subjects of the earlier studies were likely to still be growing. Another difficulty inherent in the earlier studies was demonstrated by Crile & Quiring (1939). They measured 35 Mayan soldiers in the Guatemalan Army and found that 34 of these men showed signs of goiter, a condition also endemic in the general population. Goiter, and iodine deficiency without clinical signs of goiter, are known to reduce stature. Even the most recent mean height of 169 cm (Bogin *et al.* 1992) should not be taken as definitive. This value is an estimate based on a mathematical projection of final adult height for young men measured until age 17 or 18 years. Moreover, this estimate is based on one sample from one village in Guatemala. Thus, two conclusions can be drawn regarding Mayan growth at this time: (1) Mayan men, even those who may still be growing, are taller than any 'pygmy' population; and (2) broadly representative data for Mayan adult stature in the twentieth century do not exist.

The existing research does demonstrate that Mayans in Guatemala and Mexico are shorter, on average, than Europeans of northern and western Europe and North America (Bogin, 1988a; Bogin & MacVean, 1984). In this regard Mayans are similar to other peoples of low socioeconomic status from the poor nations, who are also short-statured. An economist from the United States asserts that small body size of people who grow up in poverty is an accommodation to malnutrition (Seckler, 1980, 1982). Seckler does not discuss the Maya directly, but includes them in the category of people from poor nations. In this first paper, Seckler argues that the small size of children and adults living in poverty may be a genetic adaptation, acquired from generations of malnutrition. At first glance the Maya might seem to fit this argument. They have suffered from undernutrition, heavy workloads, and disease for the 450 years since the Conquest. Have they accommodated by evolving a smaller and less energy-demanding body? According to recent data the answer is no. As shown below in this chapter, the children of Mayan immigrants growing up in the United States average more than 5 cm greater in stature than Mayan children of the same age living in Guatemala. Clearly, if Mayan short stature were a genetic adaptation, fashioned over centuries, such a biologically and statistically significant difference in height could not occur in less than one generation.

In his second paper, Seckler rejects his earlier genetic explanation in favor of what he calls the 'homeostatic theory of growth'. By this he means that '...the single genetic potential growth curve of the older view is replaced by the concept of an *array* of potential curves in several anthropometric dimensions ... in a word, with the concept of a potential *growth space*' (1982, p. 129, author's italics). In Seckler's view, any path taken within this 'potential growth space' results in a normal healthy individual, that is, an individual without functional physical or cognitive impairment. Only in the extreme cases of under- or overnutrition does pathological growth failure or growth excess with functional impairment result. Seckler concludes '...that most of the people in the "mild to moderate" category of malnutrition are "small but healthy" people and should be considered "normal" in relation to their environment' (1982, p. 130). In this case, the 'small but healthy' category would include the Maya of Guatemala. According to Seckler, this means that nutrition interventions, although well-intended and likely to make children grow bigger, will not result in improved health or functional capacity for Mayans.

Again, at first glance this appears to be a reasonable notion. Indeed, the concept of a 'homeostatic theory of growth' correlates well with the process of developmental plasticity. However, only in the sense that homeostasis or plasticity during development results in a change in morphology is Seckler's idea correct. The assertion that growth retardation associated with undernutrition is either a genetic adaptation or a homeostatic response without functional consequence is an abuse of the concept of biological adaptation. The consensus of research with undernourished peoples of the poor nations shows that the consequences of childhood undernutrition are: (1) reduced adult body size, (2) impaired work capacity throughout life; (3) delays and permanent deficits in cognitive development; and (4) impaired school performance (Pelto & Pelto, 1989). Spurr (1983) reviews the world-wide research for physical performance and finds that '... malnutrition is accompanied by a reduced PWC [physical work capacity] ... and the degree of depression is related to the severity of the depressed nutritional status and to the loss of muscle mass' (1983, p. 21). In another review of research on the cognitive consequences of undernutrition, Pollitt & Lewis find that mild to moderate undernutrition '... affects aptitudes and abilities of pre-school children, and determine[s] in part the degree of success the child will have later within the school' (1980, p. 34). Furthermore, when such malnutrition '... is part of an economically impoverished environment, the probabilities are very high that the cognitive competencies of the child will be adversely affected'. Research in Guatemala with Mayan and *ladino* children confirms the general findings reported by Spurr and Pollitt (Bogin, 1988a, pp. 148–59; Freeman *et al.*, 1980; Martorell, 1989). Guatemalan children of low socioeconomic status lack adequate total food intake and suffer from high incidence of infectious disease, infant and child mortality, and cognitive delays and deficits. These are all indicators of the poor biological adaptation of Mayan populations. There is no evidence that reduced growth is adaptive in any sense. Undernutrition is almost always associated with poverty and, combined with the nutritional and social constraints of such poverty, results in diminished opportunities for educational, economic and sociopolitical advancement in adulthood. This situation recycles poverty, which in large part causes further malnutrition and poor growth, into future generations (Garn, Pesick & Pilkington, 1984).

The forgoing discussion of Seckler's views are presented here because they seem reasonable and, consequently, are popular both with the public and with policy makers, who might prefer to believe that undernourished children are 'small but healthy' rather than stunted and suffering. The popular view is comforting and cheaper: no economic aid is required. None the less, reasons for Mayan short stature are more likely to be associated with the social, economic, and political environments that cause poverty rather than with 'Mayan genes' or 'homeostatic adaptations'. Since the Conquest, Mayans have been politically and economically dominated by Spanish *conquistadores* or their *ladino* cultural descendants (Adams, 1970;

Handy, 1984; Smith, 1988; Warren, 1989). During the late 1970s and early 1980s the social, economic, and political fates of Guatemala's Maya deteriorated further. Anthropologist Allan Burns explains that during this time Guatemala experienced:

> ... a particularly bloody decade of civil war that has severely changed life there, especially for the Maya ... The guerrilla insurgency and the overwhelming response by the military in Guatemala resulted in the destruction of hundreds of Maya towns and villages. The Maya of the mountainous area where the guerrilla forces found refuge were caught in an uprising that left them most vulnerable: they could not quickly leave their lands and villages like the insurgents and could not defend themselves against the weaponry of the state. As Beatriz Manz (1988) has documented, the destruction of the villages and societal structures in the area has been thorough   (Burns 1989, pp. 21–22).

Thousands of Maya were killed during the civil war; more than 250 000 fled their villages and crossed the border into Mexican refugee camps. After a few years the Mexican government disbanded these squalid camps and the Maya were forced to disperse. Thousands found their way, by foot in many cases, to the United States. In the USA there are two major centers of Mayan refugees: Los Angeles, California, and Indiantown, Florida.

In these places the Maya live in relative safety and enjoy some of the nutritional and health benefits of life in the United States. The Maya migration affords the opportunity to study the consequences of life in a new environment on the health and growth of Mayan children. Such a study helps to explain the causes of Mayan short stature in Guatemala and the potential for Mayan growth in less repressive environments.

A brief history of Indiantown and its Mayan population

Indiantown, Florida, is named after a Seminole encampment, mostly obliterated today, that now serves as a community park. The town, with a permanent population of just over 3000 people, lies at the heart of the citrus and winter vegetable zone of south Florida. During the harvest season (October through May) the population almost doubles as migrant workers stream into the area. According to Allan Burns:

> Indiantown is ... located in the western part of Martin County, Florida. The county is centered on the east coast community of Stuart where the boom in Florida real estate still provides county residents with jobs and rapid population growth. To the west, in the sparsely populated agricultural zone of Indiantown, a few middle class retirement and recreational areas exist, but for the most part, communities like Indiantown remain untouched by the wealth of South Florida. Local white residents of

the town [and a few of the Blacks and Mexican-Americans] find jobs in schools, agricultural service industries and a defence contractor industry south of the community. Other residents, including American blacks, Haitians, Puerto Ricans, Mexicans, and Guatemalans have multiple occupational strategies. When farm work is available, it is sought after. When it is not, landscape work, day labor in construction, child care, salvaging and other informal sector jobs are developed (Burns 1989, p. 23).

The Maya began arriving in Indiantown in 1983 to pick oranges (Ashbranner & Conklin, 1986). More continued to come as the word spread that Indiantown provides a haven for refugees. Today the Maya are the largest of the migrant groups in town, numbering more than 1000 in permanent residents. Almost all are from the extreme northwest region of Guatemala and are speakers of the Kanjobal Maya language. When the new immigrants first arrived they lived in three apartment houses known as 'camps' because of the wretched conditions of these buildings. From 1983 to the present day, these apartments have been cited for many health and safety code violations (*The Stuart News*, Feb. 16, 1991, pp. S1–S8). Despite these conditions, rent is relatively high at about US$150.00 per week or US$525.00 per month. Often six or more single people, or two or more families, are crowded into each apartment. Today, single men and transient families still live in the apartments, but a majority of the resident families moved to trailers or houses managed by the Farm Home Administration where living conditions are improved and rents are commensurate with family earnings.

A brief history of the Los Angeles Maya

Los Angeles is often called the 'capital of the Third World' owing to the diversity of ethnic groups from developing nations living there, mostly from Latin America and Asia. United States Census data for 1990 divide the Los Angeles population of 3 485 000 people into 'Blacks' 14%, 'American Indian' 0.5%, 'Asian-Pacific Islander' 9.8%, 'Hispanic' 39.0%, with 'Whites' and all others making up the remaining 36.25%. The total population of Maya in Los Angeles is unknown, but estimates range up to 15 000.

Mayan children measured for the present research are from the Kanjobal community of Los Angeles. Thus the Indiantown and Los Angeles communities are from the same ethnic and geographic origin in Guatemala (there are also several other Mayan language or ethnic groups in Los Angeles). The first Kanjobales began arriving in Los Angeles in the early 1970s. Today there are an estimated 4000 Kanjobales in the metropolitan area. James Loucky, a social anthropologist, has worked with this community for about 10 years. He notes that: 'Most Maya over the age of 15 toil for 50 to 60 hours a week doing menial sewing work in the sweatshops of the garment district. They are underpaid and have little job security' (Loucky, 1993). A few Maya have established their own sewing shops (*maquiladores*) and employ other Maya. A few others work as health para-professionals (nurses' aids) or have semi-skilled jobs (hairdresser, electronic technician). Most of the children under age 15 years attend school. It is not known what proportion finish high school. During fieldwork in Los Angeles the present author met, or heard about, three or four of the children of these immigrants who are attending community colleges or state universities.

Most of the Kanjobales live in south-central Los Angeles, the area of the 1992 riots. Virtually all Mayans live in apartments or in rented houses often shared by two families. Rents are expensive (US$600 – 900 per month for flats and more that US$1000 per month for 'ghetto housing') in proportion to salaries, which are usually at or below minimum wage.

The Indiantown and Los Angeles samples of the present study
Children attending the two elementary schools of Indiantown were measured. One, the Warfield School, is a public school administered by the Martin County School District. Warfield has about 575 students enrolled in grades kindergarten through four and in a Head Start administered pre-school program. A 1991 survey of 625 students at the Warfield Elementary School in Indiantown found that 35% of the parents are migrant farm workers (this group includes virtually all of the Maya), 38% are laborers or store employees, 2% are professionals or supervisors (*The Stuart News*, Feb. 16, 1992). Presumably the other 25% are either unemployed or did not respond to the survey. The second elementary school is Hope Rural School, which is privately incorporated and affiliated with the Catholic Church. Hope Rural School, which has about 100 students in grades kindergarten through four, was founded to serve the children of migrant farm workers. All the children in attendance at these two schools during the week of February 24–28, 1992, were measured. In all, 555 children (275 boys and 280 girls) were measured at Warfield and 99 children (54 boys and 45 girls) at Hope Rural. The total sample includes 105 Mayan children (52 boys and 53 girls). Details of the Mayan sample are given in Table 3.1.

The Los Angeles children of this study were measured in their homes or at a toy give-away event sponsored by the IXIM cultural support group. IXIM (pronounced *ee-SHEEM*, meaning 'corn' in the Kanjobal Mayan language) is a non-profit educational and cultural organization founded in

Table 3.1. *Numbers of Mayan boys and girls by age and sex for the Indiantown, Los Angeles, San Pedro 1979–1980, and San Pedro 1989–1990 samples*

Age	Indiantown		Los Angeles		San Pedro 1978–80		San Pedro 1989–90	
	B	G	B	G	B	G	B	G
4	1	4	7	8				
5	8	9	14	6	19	7	0	1
6	11	12	7	9	17	5	15	19
7	7	8	2	8	98	76	115	94
8	6	8	7	2	68	64	126	104
9	5	5	12	6	59	49	104	103
10	5	3	6	8	50	28	70	69
11	6	4	6	9	49	23	38	35
12	3	0	6	5	36	21	59	43

1986 by Kanjobal Maya migrants to Los Angeles and by several local supporters. IXIM assists Mayan immigrants as they adjust to life in Los Angeles, provides educational opportunities, and promotes the community development and cultural expression of the Maya. The toy give-away is sponsored by IXIM each year just prior to Christmas and is one example of the service–cultural functions of the group.

A total of 136 children (69 boys and 67 girls) between the ages of 4 and 14 years were measured during the period December 18–22, 1992 (Table 3.1). No socioeconomic status information was collected for individual children or their families. Information provided by leaders of IXIM and by Dr James Loucky, who know most of the families, indicate that virtually all of the children and their families conform to the general description of the Los Angeles Maya discussed above.

Measurement protocols

The measurements taken on each child were height, weight, arm circumference, and triceps skinfold. Height was measured with a Harpenden Portable Stadiometer. The child stood, without shoes, on the foot plate and the head plate was placed on the top most part of the skull with the head held in the Frankfurt plane. Measurements were read to the nearest 0.5 cm while the observer was seated or standing with his eyes level with the reading mark. Weight was recorded with a Seca portable scale. The child stood on the foot pad, wearing minimal indoor clothing, and the reading was taken to the nearest 0.5 kg. The scale was 'zeroed' at the start and several times during the measurement session. Arm circumference was measured with a steel tape at the midpoint of the upper arm between the olecranon and acromion processes. Measurements were read to the nearest millimeter. Triceps skinfold was taken with a Lange calliper at the midpoint of the upper arm, above the triceps muscle. The thickness of a double fold of skin and subcutaneous fat at this site was estimated to the nearest 0.5 mm. The present author measured all of the children. At the Warfield School a teacher's aid or administration staff member who lived in Indiantown assisted with ethnic identification of each child and recorded all the measurements. At Hope Rural the principal provided ethnic identification and a student recorded the measurements. In Los Angeles several members of IXIM assisted with intake data (name, sex, birth date, place of birth of the child) and verified the ethnicity of each child and his or her parents. An assistant recorded the anthropometric measurements.

The rationale for each of these measurements is as follows. Because height increases over time, it is an indicator of the total nutrition and health history of a child. In contrast, weight can both increase and decrease over time and, therefore, relates more to recent nutrition and health status (Waterlow *et al.*, 1977; Habicht, Yarbrough & Martorel, 1979). Circumferences and skinfolds are generally accepted measures for body composition, that is, lean body mass and fat mass. Body composition is often used as a proxy for nutritional status. Lean body mass is an indicator for the body's reserves of protein; fat mass is an indicator of the body's reserves of energy (Martorell *et al.*, 1976; Frisancho, 1981). In the present study, arm circumference and triceps skinfold were used to calculate fat area and muscle area at the midpoint of the arm (the formulae of Gurney & Jelliffe (1973) were used). These calculated areas have been shown to correlate more closely with fat and lean body mass than the raw circumferences and diameters (Frisancho, 1981, 1990).

In order to compare the growth of Maya living in the United States with Maya living in their homeland, two samples of Guatemala-resident Maya are included in the analysis. The Guatemalan Maya are samples of children between the ages of 5 and 14 years attending a public school in the village of San Pedro Sacatapequez. The first sample of these children are 753 boys and girls measured in 1979 and 1980. Previous analysis shows that they are representative in height, weight and other physical dimensions of Mayan children of low socioeconomic status (SES) in Guatemala (Bogin & MacVean, 1984). The second sample are 1154 boys and girls measured in 1989 and 1990. Analysis of these two Guatemala-resident samples provides the opportunity to account for changes in growth within one village over a decade of time. Change between Guatemalan-living samples measured over a decade can then be used to help evaluate the causes of change in growth between samples in Guatemala and the United States.

These low-SES Mayan children of Guatemala live in a village with no safe supply for drinking water, an irregular supply of water, and unsanitary means for waste disposal. The parents of the Mayan children are employed, predominately, as tailors or seamstresses by local clothing manufacturers or are self-employed street vendors of textiles and agricultural produce. Textile workers in this town are usually paid 'piecework' wages, e.g. about US$0.25 for a finished blouse or shirt. There is one public health clinic in the village, and the treatment of infants and preschool children with clinical undernutrition is common. Body composition analysis of the children attending the public school finds that, on average, they exhibit signs of chronic, mild to moderate undernutrition (Bogin & MacVean, 1984).

Results

The first question to be answered from these data is: How do Mayan children living in Indiantown and Los Angeles compare with Mayans living in Guatemala? The second question is: How do the Mayan children of Indiantown compare with children of other ethnic groups living in the United States? To answer both questions the data for girls and boys were combined. Combining the sexes is justified by multiple regression analysis of the effects of AGE, SAMPLE (Indiantown, Los Angeles, or Guatemala) and SEX (the independent variables) on each of the growth measurements (dependent variables), with a significance level set at $p = 0.01$. SEX does not have a significant effect on height, weight, arm circumference, triceps skinfold, or fat area, but does have a significant effect on muscle area; boys have more muscle. This body composition differences in children is well known, and although statistically significant, it is small prior to puberty. Since all the samples show the same magnitude of difference, i.e., no statistical interaction, combining data for boys and girls within groups does not change the effect of AGE or SAMPLE on the growth data. Moreover, merging the data for boys and girls increases sample sizes, thus increasing the power of the statistical tests, and simplifies the analysis.

The results of the multiple regression testing also allow for a further simplification of the data. There are no significant differences between the Indiantown and Los Angeles samples of Mayan children for any of the anthropometric variables. Thus these two samples are combined for analysis, and are referred to in the rest of this chapter as the LA–IT sample. As an aside, it is worth noting that the Indiantown sample is from a rural area and the Los Angeles sample is from an urban area. In the economically underdeveloped nations of the world, such as Guatemala, there are

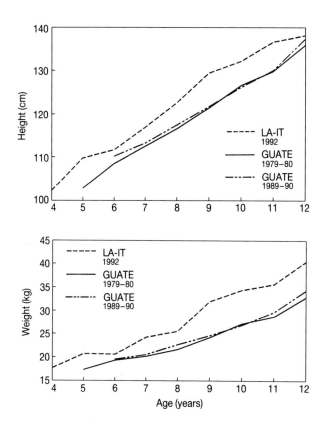

Fig. 3.1. Comparisons of mean height or weight of Los Angeles and Indiantown Maya (LA–IT) with San Pedro, Guatemala, Maya (GUATE) children by age. Data for boys and girls within samples are combined.

significant difference in child growth between urban centers and rural areas. However, in economically developed nations, such as the United States and the United Kingdom, there are no longer any rural–urban differences in the growth of the resident population (Bogin, 1988a, p. 156). Even the Mayan children of the present study, who have only recently migrated, show no rural–urban effect on growth.

Growth status of United States versus Guatemalan Mayans
The mean value for height and weight at each age (ascertained from birth certificates or official school records) for the LA–IT sample and the 1979–80 sample from Guatemala (abbreviated S.P. for the village name San Pedro) is presented in Fig. 3.1. In Fig. 3.2 and 3.3 are presented similar

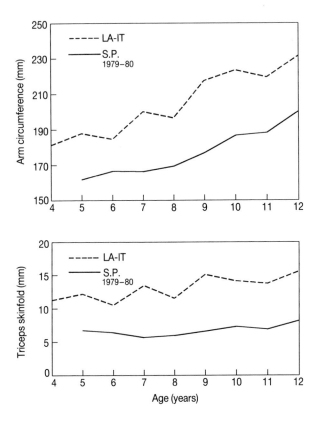

Fig. 3.2. Comparison of mean arm circumference or triceps skinfold of Los Angeles and Indiantown Maya (LA–IT) with San Pedro, Guatemala, Maya (S.P.) children by age. Data for boys and girls within samples are combined.

comparisons for the other growth variables. Children of the LA-IT sample range in age from 4 to 14 years. However, there are too few children over age 12 for meaningful analysis. In each figure, the mean values for the LA–IT sample are significantly greater than those of the S.P. samples. For height, the average difference between samples is 5.5 cm. Fig. 3.1 also includes mean values for height and weight of the S.P. sample measured in 1989–90. There are no significant differences between the means of the older and more recent S.P. samples.

The answer to the first question is straightforward: the Mayan children of Indiantown are significantly taller, heavier, and better nourished than Mayan children in Guatemala.

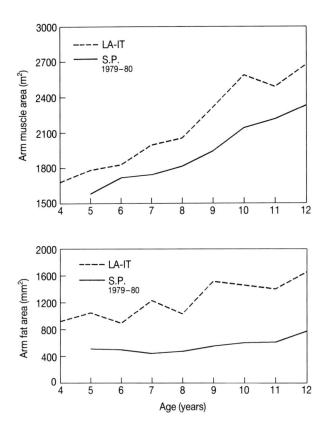

Fig. 3.3. Comparison of mean arm muscle area or arm fat area of Los Angeles and Indiantown Maya (LA–IT) and San Pedro, Guatemala, Maya (S.P.) children by age. Data for boys and girls within samples are combined.

Growth status of Mayans compared with other ethnic groups living in the United States

In Indiantown all of the children attending the two elementary schools were measured. This permits a comparison of Mayan children with children from other ethnic groups resident in the same town. Ethnic classification of the children of Indiantown is determined in three ways. The first is self-reported ethnicity by the parents of the children, for instance on school or government forms. The second way is determination by one or more residents of the town; for example, the teacher's aid and the secretary assisting with the measurements at the Warfield school were familiar with most of the children and their family background. The third method was to ask the child his or her ethnicity. Only the youngest children had difficulty

Table 3.2. *Numbers of boys and girls measured in Indiantown in each ethnic group by age*

Age	Mayan		Mexican		White		Black	
	B	G	B	G	B	G	B	G
4	1	4	6	3	0	1	5	3
5	8	9	17	8	11	4	11	14
6	11	12	17	20	10	9	10	18
7	7	8	17	16	11	11	9	21
8	6	8	9	11	16	13	19	16
9	5	5	16	10	14	8	11	16
10	5	3	9	12	9	10	9	11
11	6	4	3	3	0	0	4	2
12	3	0	0	0	0	0	0	0

with this question. The following ethnic groups were identified in this manner: Maya, Mexican-American (including citizens of both Mexico and the United States), Haitian, Puerto Rican, White, Black, White/Mexican, Jamaican, and Filipino. Not all of these ethnic groups had enough children for meaningful statistical analysis. Four groups did have sufficient numbers. These are: Mayan (52 boys, 54 girls), Mexican (95 boys, 82 girls), White (71 boys, 56 girls), and Black (78 boys, 101 girls). Samples sizes of each ethnic group, by sex and age, are presented in Table 3.2.

The mean values by age and ethnicity for each of the growth variables are presented in Fig. 3.4 and 3.5. For height (Fig. 3.4) it is clear that Mayan children are, on average, the shortest. Indeed, analysis of variance with Scheffé post-hoc contrasts of means shows that the Mayan ethnic group is significantly shorter than each of the other ethnic groups. Mexican-Americans are shorter than Blacks or Whites, but there is no statistical difference between Blacks and Whites. Mean values of height for a national sample of United States children (NCHS) are included in the comparisons for height. The White and Black samples of Indiantown are comparable in height to the national sample. This is true for each of the other anthropometric variables as well.

Mayans, as a group, weigh significantly less than Whites or Blacks. There is no difference between Mayans and Mexican-Americans, nor are there differences between the White and Black ethnic groups. There are no ethnic differences in body composition measures, such as for arm fat area or for arm muscle area (Fig. 3.5).

Thus, the answer to the second question is: Mayan children of the LA–IT sample are, on average, shorter and weigh less than children of the other

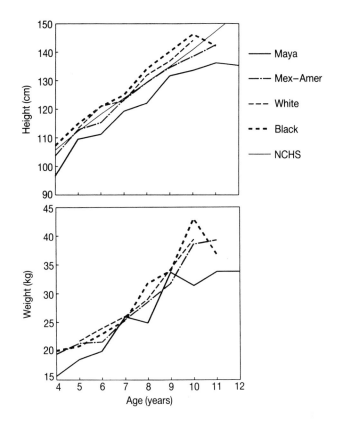

Fig. 3.4. Mean height or weight of Indiantown children by ethnic group and age.

ethnic groups. However, in terms of body composition the Mayan children, as a group are equal to the other ethnic groups. Consequently, for their smaller stature and weight, the Mayan children carry relatively more muscle and fat than children of the other ethnic groups.

Discussion
The present research examines the first generation of Mayan migrants to the United States. Their growth in height, weight and several measures of body composition is significantly greater than that of Mayan children in Guatemala, but less than that of long-term residents of the United States. Height, weight, and body composition measurements (arm circumference, triceps skinfold, fat area and muscle area) are indirect indicators of dietary intake, nutrient utilization, and energy expenditure of the children.

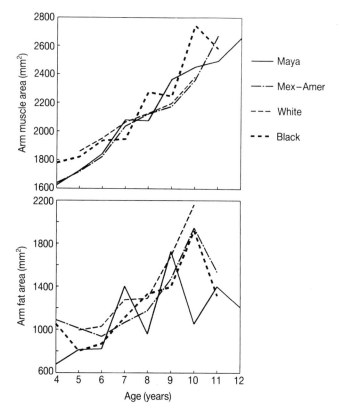

Fig. 3.5. Mean arm muscle area or arm fat area of Indiantown children by ethnic group and age.

Although indirect, these measures correlate well with direct observation of food intake, nutrient utilization, episodes of severe illness, and energy expenditure in work, play and illness (Habicht et al., 1979; Frisancho, 1990). Consequently, growth measurements are widely used by public health workers, nutritionists, and anthropologists to characterize the nutrition and health status of populations.

Children of the long-term resident ethnic groups (Black, Mexican-American, and White) in Indiantown are similar in height, weight, and body composition to national samples of children from the same ethnic groups published by the National Center for Health Statistics (Hamill et al., 1977; Johnson et al., 1981; Najjar & Kuczmarski, 1989). The body composition of the LA–IT Maya sample is, on average, similar to that of the other ethnic groups measured in Indiantown. By these criteria the Mayan children of Indiantown, Florida, and Los Angeles, California, appear to be generally healthy and well nourished. However, the average height of the Mayans is significantly less than that of all other ethnic groups, and the average weight of the Mayans is significantly less than that of the Blacks and Whites. Since there is no evidence that the Mayans are undernourished, their lower average weight is probably a consequence of their shorter average stature. The reason for that shorter stature, in the face of what appears to be a good environment for health and nutrition is, of course, the central focus of this study.

Determinants of growth

Clearly, the short stature of Indiantown Mayan children, relative to the other ethnic groups of Indiantown, is not due to simple genetics. If this were true, then the Mayans living in the United States would be similar in stature to the Mayan children living in the Guatemalan village of San Pedro. However, the LA–IT sample is significantly taller than the older and more recent San Pedro samples. A review of the literature finds that the LA–IT sample is, on average, taller than any sample of Guatemalan Mayans and taller than many samples of low-SES *ladinos* of Guatemala (Bogin & MacVean, 1984; Bogin *et al.*, 1989; Johnston *et al.*, 1985; Plattner, 1974). Thus, the improvement in the nutritional and health environment, from Guatemala and the refugee camps in Mexico to Indiantown and Los Angeles, results in a significantly greater average stature in the Mayan children. But, if the Maya are equal to other ethnic groups in body composition, why are they still shorter?

Secular trends in growth

The most likely answer is that the present generation of Mayan children living in the United States are in the first stage of a process of increasing stature from generation to generation. This process, known as the secular trend in growth, is a well-known event, often associated with migration from a worse to a better environment (Bogin, 1988b). This process is called the secular trend because it is found in the growth of migrant children from all social, economic, religious, and ethnic groups. Some of the most important examples of the secular trend are the publications of Franz Boas on European children (1912, 1940), Shapiro on Japanese children (1939), and Goldstein (1943) and Lasker (1952) on Mexican children that were discussed above. Moreover, follow-up studies of these same populations show that over time the growth in height of each generation of children continues to increase (Roche, 1979).

In most studies of secular trends the increase in mean height from generation to generation lags behind increases in weight and body composition. This happens because, as explained earlier, height reflects health and nutritional history, whereas weight and body composition reflect recent events. Indeed a child's height is an historical record of both the individual and his or her parents. This is due to cross-generational effects of chronic undernutrition and disease. These insults during childhood are known to reduce the growth of future offspring (Van Wieringen, 1986). Conversely, children who are better nourished and healthier will give their own offspring a healthier prenatal start in life. Certainly, bigger mothers have longer, heavier babies who grow up to be taller children and adults (Garn et al., 1984). One example of this phenomenon comes from Mexican immigrants to the United States. A review of cross-generational studies of growth demonstrates that Mexican-American children have become taller, on average, with each generation since the 1930s (Bogin, 1989).

The Maya of Indiantown and Los Angeles conform to the model of migration and growth described in these other studies. Indeed, the Mayan biological response to migration exceeds that of the earlier cases. Boas reported mean changes in stature of immigrants to the United States that ranged from -1.8 cm for Neapolitan girls to $+2.9$ cm for Bohemian boys (Boas, 1912). The increase in size between the Mayan children now living in the United States and Mayans in Guatemala amounts to an average of 5.5 cm in stature and 4.7 kg in weight, and proportionately large increases in the other measures of body composition as well, across all ages. This magnitude of secular change is found in studies of other populations that change from traditional rural to modern environments. For example, the children of Samoans who migrate from rural villages to cities in American Samoa or Hawaii show the typical response of increased growth. The last decade has also brought the modern world to the rural villages of Western Samoa. Children living in Western Samoa were measured in 1979–82 and again in 1991 (Asnis & McGarvey, 1993). The most recent sample of pre-teenagers averages '. . . almost 4 kg heavier and more than 5 cm taller' than then earlier sample. 'These differences [are] attributed to the dietary, activity level and health care changes of modernization' (1993, p. 51). This Samoan study shows, once again, that it is the environment for nutrition and health that brings about the response in growth, rather than any genetic change or even the act of migration itself.

Archaeological evidence for Mayan stature

Another source of information concerning secular changes in growth for Mayans comes from archaeological excavations of Classic and Post-Classic tombs. These studies, so far confined to male skeletal remains, show that

Mayan average stature was greater in the past than it is today (Haviland, 1967). During the Early Classic period, AD 250–400 (i.e. the beginning of state society formation) at the site of Tikal the stature of skeletons from tombs, an indication of high social status, averages 172 cm. Moreover, Haviland reports that these tomb skeletons have a '...marked robust appearance' (1967, p. 320), which is interpreted as a sign of good health and nutritional status. In contrast, non-tomb skeletons have an average stature of 164 cm, a difference that Haviland attributes to the lower social status of these commoners, and their lower nutrition and health status. During the Late Classic period, AD 700–900, the average stature of skeletons from both tomb and non-tomb burials decreases to 162 cm and 156 cm, respectively. These values are comparable to the stature of post-Conquest Mayan populations measured in this century.

A variety of causes are cited to explain the decrease in average stature during the Classic period. Haviland states that 'By the Late Classic times, population at Tikal seems to have become sufficiently heavy to place a severe strain on the agricultural potential of the region' (1967, p. 323). Other researchers add that this agricultural potential may have been compromised by degradation of the farmlands and food-producing systems, diseases to crops and the dense human population, fragility of the increasingly oppressive political system, and the possibility of sudden and violent internal revolt (Webster, Evans & Sanders, 1993, p. 533).

Recent Mayan history and growth
Today, a thousand years after the collapse of Classic Maya civilization, many of these same reasons may be offered to explain the low socioeconomic status, poor health status, and short stature of contemporary Mayan populations. The Tojalabal-speaking Maya of Chiapas Mexico, for example, have been studied in terms of demography and child growth (Furbee *et al.*, 1988). It is known that these rural agriculturalists are experiencing extreme degrees of socioeconomic and sociopolitical stress. In terms of growth, Tojolabal Maya between the ages of 6 and 15 years are shorter, lighter, smaller in head circumference, and, generally, smaller in triceps skinfolds than any other Mexican population. The Tojolabal Maya average more than 5 cm shorter at all ages than the Kaqchiquel-speaking Maya of San Pedro, Guatemala. Lack of basic human services and insufficient land for food production in the Tojolabal villages are reflected by the poor growth of the children.

Despite being 5 cm taller, on average, than the Tojolabal Maya children, the children of San Pedro village also live in an environment of chronic undernutrition. Growth data indicate that this situation has not improved since the early 1960s (Bogin & MacVean, 1984). Indeed, surveys taken during the 1980s indicate that health and nutritional status of the rural Mayan population in Guatemala had declined over the previous 20 years (Bossert & Peralta, 1987). As shown in Fig. 3.1, there is little difference in size between the children measured in 1979–80 and 1989–90. By negative association, the deterioration of an already poor environment for child growth and the absence of change over time in growth for the children of San Pedro is strong evidence for an environmental interpretation of the patterns of growth of Mayan children. Moreover, descriptions of life after the civil war of the late 1970s, both in Guatemala and in the Mexican refugee camps, portray dreadful conditions for child development (Manz, 1988). It is known that living conditions of the Maya of Indiantown and of many Maya in Los Angeles have improved considerably since they first began arriving in 1983 (Ashbranner & Conklin, 1986; Loucky & Harmon, 1993). Thus, the values of height for Mayan children in the United States, intermediate between averages for Guatemala-living Maya and long-term residents of the United States, do seem to reflect the history of change in the environments for growth. In conclusion, it is likely that the increase in stature of Mayans living in the United States represents a return to physical size that was common more than a thousand years ago.

Plasticity, time, and growth

Additional support for this conclusion comes from an analysis of time and growth for the United States-living Maya. Recall that in their work on plasticity, Boas, Lasker, and Lasker & Evans also found that the longer the time of exposure to the new environment the greater the effect on growth. A time-dependent effect is also evident in the present study. At the Hope Rural School, 29 of the 52 Mayan children were born in Guatemala, whereas the others were born in the United States. Place of birth is not known for the Maya attending the Warfield School. In the Los Angeles sample, 74 of the 140 children were born in Guatemala or the refugee camps of Mexico. All of the parents of the children were born in Guatemala. The mean height of those children born in the United States, adjusted by analysis of covariance for differences in age, is 123.6 cm. The mean adjusted height for the children born in Guatemala and Mexico is 121.6 cm. The difference is statistically significant at the $p = 0.10$ level. Although this difference is small, the trend for growth is in the direction predicted by the work of Boas, Lasker and their followers. Indeed, Mayan children living in the United States provide a nearly ideal example of the environmental model of plasticity of growth.

Plasticity and adaptation

Schell (Chapter 11) differentiates between a *plastic response* to environmental change and a *plastic adaptation* to an environmental stress. A descriptive growth study may document a plastic response. Evidence for a plastic adaptation requires documentation of irreversible changes in the adult phenotype and analysis of the consequences for survival, functional capacity, and reproduction (Lasker, 1969; Scrimshaw & Young, 1989).

The Mayan children of the present study certainly show a plastic response in their growth to the environment. By virtue of their increase in size over Mayans living in Guatemala the Mayan children growing up in the United States already show evidence of improved health. Not because 'bigger is better' in any absolute sense, but because increased size is the result of more nutrients for growth rather than for work, repair of the body, or combating disease. In this sense the plastic response in the growth of the Mayan children is a proxy for the quality of the environment in which they live. In Guatemala the quality of the environment is low and growth is depressed; in the United States the environmental quality is better and growth is promoted.

It is not possible to state if there is any evidence of a plastic adaptation in Mayan growth, either in Guatemala or in the United States. Longitudinal measurement and analysis of the samples described here are required for this, including information on irreversible changes in their adult phenotypes, survival to adulthood, adult functional capacity, and completed fertility. Even if such data are forthcoming their interpretation may be rather difficult. It is likely that Mayan children raised in the United States will be taller than their parents or Mayans growing up in Guatemala. It is also likely, since they are healthier, that Mayans living in the United States will have lower mortality at all ages compared with Mayans in Guatemala. Functional capacity may increase, in general, for the Mayans raised in the United States. Larger body size and increased lean body mass are positively associated with worker productivity (Spurr, 1983; Stinson, 1992) and with cognitive function in some realms (Pollitt & Lewis, 1980; Martorell, 1989). Completed fertility, however, of the United States Mayans may be lower than that of the Guatemalan Mayans. This is because fertility is inversely related to rural-to-urban migration, years of education, socioeconomic advancement from lower to middle class, and other related changes in living conditions (Bogin, 1988b). Figure 3.6 illustrates this tendency for both White and Hispanic women in the United States according to years of education. The data include women of all marital classes between the ages of 35 and 44 years old. Note that by 16 years of education the fertility of

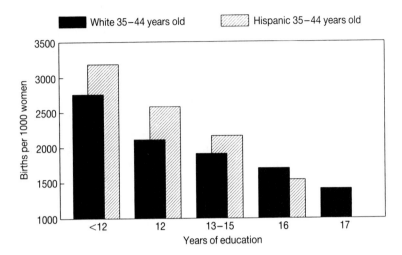

Fig. 3.6. Births to White and Hispanic women living in the United States (source: US Bureau of the Census, Current Population Reports, Series P-20, No. 427, *Fertility of American Women*: *June 1987*, US Government Printing Office, Washington, DC, 1988).

Hispanic women falls below that of Whites, showing that the education effect can be stronger than the ethnic effect. The Hispanic women described here are of Mexican, Puerto Rican, and Cuban origin. There is no reason to expect that Guatemalan Mayan women will behave any differently. How is one to decide, then, if a plastic adaptation has occurred if United States living Mayans survive longer, end up taller, increase functional capacity, but have lower completed fertility than their Guatemalan counterparts? Is the attainment of three of the four currently used criteria sufficient evidence for adaptation?

Conclusion
Mayans in the United States show a plastic response, but it is not known whether they display a plastic adaptation to their new environment. The Mayan children of Indiantown and Los Angeles are taller, heavier, and carry more fat and muscle mass than Mayan children living in an impoverished village in Guatemala. However, they are shorter, on average, than children of other ethnicities (White, Black, and Mexican-American) living in Indiantown. It is likely that the Mayan migrants to the United States are in the process of the secular trend in height growth. It is expected that each succeeding generation of Mayan children born in the United States will show an increase in stature, just as has occurred with

Mexican-American children over time. A major reason for the secular trend is the relief from the chronic undernutrition and disease that virtually all living Mayan adults experienced during their childhood. Thus, within a few generations, United States-born children of Mayan descent, if allowed equal access to health, economic, and social opportunities, should have an average stature similar to that of Americans as a whole.

Acknowledgements
Sincere thanks to Sister Carol Jean, Sister Marie Celeste, and Mr Antonio Silvestre of the Hope Rural School, and to Mr Raymond Parrish, Principal of Warfield Elementary School in Indiantown, Florida. Ms Pansy Leonard and Ms Christina Garcia assisted with the measurements of children at the Warfield School. Miss Sarah Banks assisted with measurements at Hope Rural School. Dr James Loucky arranged for research to be conducted in Los Angeles, California. While in Los Angeles the author was assisted by Mr Juan Gaspar, Mr Miguel Mendez, and Dr Fernando Peñalosa of the IXIM Cultural Center. Mr Horacio Mendez, Mr Matias Pasqual and Mrs Petrona Jimenez assisted with the measurements. Professors Derek Roberts, Gabriel Lasker, Larry Schell, and Ms Mary Ann Wnetrzak offered helpful criticisms of early drafts of this chapter. Financial support for this project was provided by the Timothy and Jean Morbach Research Fund of the University of Michigan-Dearborn.

References
Adams, R. N. (1970) *Crucifixion by Power*. Austin: University of Texas Press.
Ashbranner, B. & Conklin, P. (1986). *Children of the Maya*. New York: Dodd, Mead.
Asnis, L. J. & McGarvey, S. T. (1993). Child growth in Western Samoa in 1991 (abstract). *American Journal of Physical Anthropology*, Supplement **16**, 51.
Boas, F. (1910). *Changes in the Bodily Form of Descendants of Immigrants* (abstract). (Report submitted to the Congress of the United States of America.) New York: Columbia University.
Boas, F. (1912). Changes in the bodily form of descendants of immigrants. *American Anthropologist* **14**, 530–3.
Boas, F. (1922). Report on the anthropometric investigation of the population of the United States. *Journal of the Amerian Statistical Association* **18**, 181–209.
Boas, F. (ed.) (1940). *Race, Language and Culture*. New York: Free Press.
Bogin, B. (1988a). *Patterns of Human Growth*. Cambridge University Press.
Bogin, B. (1988b). Rural-to-urban migration. In: *Biological Aspects of Human Migration*, ed. C. G. N. Mascie-Taylor & G. W. Lasker, pp. 90–129. Cambridge University Press.
Bogin, B. (1989). Biological effects of urban migration on Hispanic populations (abstract). *American Journal of Physical Anthropology* **78**, 194.
Bogin, B. & MacVean, R. B. (1984). Growth status of non-agrarian, semi-urban living Indians in Guatemala. *Human Biology* **56**, 527–38.

Bogin, B., Sullivan, T., Hauspie, R. & MacVean, R. B. (1989). Longitudinal growth in height, weight, and bone age of Guatemala Ladino and Indian Schoolchildren. *American Journal of Human Biology* **1**, 103–13.

Bogin, B., Wall, M. & MacVean, R. B. (1992). Longitudinal analysis of adolescent growth of *ladino* and Mayan school children in Guatemala: effects of environment and sex. *American Journal of Physical Anthropology* **89**, 447–57.

Bossert, T. J. & Peralta, E. del Cid (1987). *Guatemala Health Sector Assessment 1987 Update*. Guatemala: USAID Mission in Guatemala.

Burns, A. (1989). The Maya of Florida. *Migration World* **27**(3/4), 20–6.

Crile, G. W. & Quiring, D. P. (1939). A study of metabolism of the Maya-Quiché Indian. *Journal of Nutrition* **18**, 369–74.

Diamond, J. (1992). A question of size. *Discover* **13**(5), 70–7.

Freeman, H. E., Klein, R. E., Townsend, J. W. & Lechtig, A. L. (1980). Nutrition and cognitive development among rural Guatemalan children. *American Journal of Public Health* **70**, 1277–85.

Frisancho, A. R. (1981). New norms of upper limb fat and muscle areas for the assessment of nutritional status. *American Journal of Clinical Nutrition* **34**, 2450–5.

Frisancho, A. R. (1990). *Anthropometric Standards for the Assessment of Growth and Nutritional Status*. Ann Arbor: University of Michigan Press.

Furbee, L., Thomas, J. S., Lynch, H. K. & Benfer, R. A. (1988). Tojolabal Maya population response to stress. *Geoscience & Man* **26**, 17–27.

Garn, S. M., Pesick, S. D. & Pilkington, J. J. (1984). The interaction between prenatal and socioeconomic effects on growth and development in childhood. In: *Human Growth and Development*, ed. J. Borms, R. Hauspie, A. Sand, C. Susanne, & M. Hebbelinck, pp. 59–70. New York: Plenum.

Goff, C. W. (1948). Anthropometry of a Mam-speaking group of Indians from Guatemala. *American Journal of Physical Anthropology* **6**, 429–48.

Goldstein, M. S. (1943). *Demographic and Bodily Changes in the Descendants of Mexican Immigrants*. Austin: Institute of Latin American Studies.

Gurney, T. M. & Jelliffe, D. B. (1973). Arm anthropometry in nutritional assessment: nomogram for the rapid calculation of muscle circumference and cross-sectional muscle and fat areas. *American Journal of Clinical Nutrition* **26**, 912–15.

Habicht, J. P., Yarbrough, C. & Martorell, R. (1979). Anthropometric field methods: criteria for selection. In: *Human Nutrition: A Comprehensive Treatise*, vol. 2, ed. R. B. Alfin-Slater & D. Kritchevsky, pp. 365–87. New York: Plenum.

Hamill, P. V., Johnson, C. L., Reed, R. B. & Roche, A. F. (1977). NCHS growth curves for children. National Center for Health Statistics. *Vital and Health Statistics*, **11**(165).

Handy, J. (1984). *Gift of the Devil: A History of Guatemala*. Boston: South End Press.

Haviland, W. A. (1967). Stature at Tikal Guatemala: implications for ancient Maya demography and social organization. *American Antiquity* **32**, 316–25.

Johnson, C. L., Fulwood, R., Abraham, S. & Bryner, J. D. (1981). Basic data on anthropometric measurements and angular measurements of the hip and knee joints for selected age groups 1–74 years of age. National Center for Health Statistics. *Vital and Health Statistics*, **11**(219).

Johnston, F. E., Low, S. M., Baessa, Y. de, & MacVean, R. B. (1985). Growth status of disadvantaged urban Guatemalan children of a resettled community. *American Journal of Physical Anthropology* **68**, 215–24.

Lasker, G. W. (1952). Environmental growth factors and selective migration. *Human Biology* **24**, 262–89.

Lasker, G. W. (1969). Human biological adaptability. *Science* **166**, 1480–6.

Lasker, G. W. (1984). The morphology of human populations. In: *Estudios de Antropologia Biologica II: Coloquio de Antropologia Fisica Juan Comas, 1982*, ed. R. Galavan, R. Ramos Rodriguez & R. Ma. Ramos Rodriguez, pp. 145–57. Mexico: Universidad Nacional Autonoma de Mexico.

Lasker, G. W. & Evans, F. G. (1961). Age, environment and migration: further anthropometric findings on migrant and non-migrant Mexicans. *American Journal of Physical Anthropology* **19**, 203–11.

Loucky, J. (1993). Central American Refugees: learning new skills in the U.S.A. In: *Contemporary Anthropology*, ed. M. C. Howard (4th edn), pp. 228–30. New York: HarperCollins.

Loucky, J. L. & Harmon, R. (1993). Where do we go from here? Guatemalan Mayas and urban violence in Los Angeles. Paper presented at the American Association for the Advancement of Science meetings, Boston, Massachusetts.

Manz, B. (1988). *Refugees of a Hidden War: The Aftermath of Counterinsurgency in Guatemala*. Albany: State University of New York Press.

Martorell, R. (1989) Body size, adaptation, and function. *Human Organization* **48**, 15–20.

Martorell, R., Yarbrough, C., Lechtig, A., Delgado, H. & Klein, R. E. (1976). Upper arm anthropometric indicators of nutritional status. *American Journal of Clinical Nutrition* **29**, 46–53.

Méndez, J. & Behrhorst, C. (1963). The antropometric characteristics of Indian and urban Guatemalans. *Human Biology* **35**, 457–69.

Najjar, M. F. & Kuczmarski, R. J. (1989). Anthropometric data and the prevalence of overweight for Hispanics: 1982–84. National Center for Health Statistics. *Vital Health Statistics* **11**(239).

Pelto, G. H. & Pelto, P. J. (1989). Small but healthy? An anthropological perspective. *Human Organization* **48**, 11–15.

Pollitt, E. & Lewis, N. (1980). Nutrition and educational achievement. Part 1, Malnutrition and behavioral test indicators. *Food and Nutrition Bulletin* **2**, 32–4.

Plattner, S. (1974). Wealth and growth among Mayan Indian peasants. *Human Ecology* **2**, 75–87.

Roche, A. F. (ed.) (1979). Secular trends in human growth, maturation, and development. *Monographs of the Society for Research in Child Development* **44**, no. 179.

Scrimshaw, N. S. & Young, V. R. (1989). Adaptation to low protein and energy intakes. *Human Organization* **48**, 20–30.

Seckler, D. (1980). Malnutrition: an intellectual odyssey. *Western Journal of Agricultural Economics* **5**, 219–27.

Seckler, D. (1982). "Small but healthy": a basic hypothesis in the theory, measurement, and policy of malnutrition. In: *Newer Concepts of Nutrition and their Implication for Policy*, ed. P. V. Sukhatme, pp. 127–37. Pune, India: Maharashta Association for the Cultivation of Science Research Institute.

Shapiro, H. L. (1939). *Migration and Environment*. Oxford University Press.

Shattuck, D. C. & Benedict, F. G. (1931). Further studies of the basal metabolism of Maya Indians. *American Journal of Physiology* **96**, 518–28.

Smith, C. A. (1988). Destruction of the material bases for Indian culture: Economic changes in Totonicapán. In: *Harvest of Violence: The Maya Indians and the Guatemalan Crisis*, ed. R. M. Carmack, pp. 206–31. Norman: University of Oklahoma Press.

Spurr, G. B. (1983). Nutritional status and physical work capacity. *Yearbook of Physical Anthropology* **26**, 1–35.

Steggerda, M. & Benedict, F. G. (1932). Metabolism in Yucatan: a study of the Maya Indian. *American Journal of Physiology* **100**, 274–84.

Stinson, S. (1992). Nutritional adaptation. *Annual Review of Anthropology* **21**, 143–70.

Van Wieringen, J. C. (1986). Secular growth changes. In: *Human Growth: A Comprehensive Treatise* (2nd edn), vol. 3, ed. F. Falkner & J. M. Tanner, pp. 307–31. New York: Plenum.

Warren, K. B. (1989). *The Symbolism of Subordination: Indian Identity in a Guatemalan Town* (2nd edn). Austin: University of Texas Press.

Waterlow, J. C., Buzina, R., Keller, W., Lane, J. M., Nichaman, M. Z. & Tanner, J. M. (1977). The presentation and use of height and weight data for comparing the nutritional status of groups of children under the age of 10 years. *Bulletin of the World Health Organization* **55**, 489–98.

Webster, D. L., Evans, S. T. & Sanders, W. T. (1993). *Out of the Past*. Mountain View, California: Mayfield.

4 The place of plasticity in the study of the secular trend for male stature: an analysis of Danish biological population history

JESPER L. BOLDSEN

Summary

Adaptation implies changes in the human phenotype induced through interaction with the environment. There are four different levels of adaptation: acclimatization, plasticity, population structure and natural selection. The different levels of adaptation are active on different timescales. The present analysis concentrates on effects of plasticity and population structure using data on the stature of Danish males from archaeological and written historical sources. These levels of adaptation operate on a timescale of individual lifespan and across several generations of individuals.

The genealogical relations among the men forming the material cannot be specified. Consequently, data cannot be analysed along the lines of ordinary quantitative genetics. The focus of the analysis must be changed from individuals to groups of individuals. In this change the regression analysis is changed to an analysis of variance (ANOVA). Two groups are defined based on the conditions of living and growing; two other groups are based on the population structure in the communities from which the analysed samples were taken. The first two groups have been classified as either permissive (i.e. living in a community with good conditions) or restrictive. The second set of groups are classified as either inbred or outbred.

The stature of a mother has been shown to affect the stature of her offspring in a nonlinear way. This maternal effect is likely to postpone the effect of changes of the population structure and/or the conditions of living on stature by at least one generation. Consequently, only data extracted from populations assumed to have lived under fairly constant conditions and with a constant structure for several generations are analysed. Data have been drawn from measurements of stature in the graves of two mediaeval cemeteries and from the medical examination of young men prior to conscription. Populations could be classified in four categories according to conditions of living and population structure; but in practice

75

only three categories are available. It has not been possible to find an inbred community living in a permissive environment.

During the past 140 years, mean male stature in Denmark has increased by some 13 cm. The analysis indicate that the population structure effect on stature was 5.9 cm and that the plasticity effect was 7.2 cm. Thus, 45% of this increase is due to the population structure effect and the remaining 55% is due to improved conditions of living and growing.

Introduction

The concept of adaptation implies environmentally induced change from one set of biological stages to another on at least one of several mutually interacting levels of organization. Natural selection modifies the gene frequencies at the population or species level. The population (breeding) structure modifies the genotype frequencies and may interact with natural selection by exposing different sets of genotypes to the forces of natural selection. Ontogenetic adaptation (plasticity) modifies the permanent (adult) phenotype of the individual. This process interacts both with the population structure (through mate-selection) and with natural selection (by exposing genetic differences in vulnerability). Physiological and behavioural response to environmental conditions (acclimatization) is active on the same level as plasticity but it does not leave a permenent mark on the phenotype of the individual. In this chapter the term phenotype is used in a broader sense than usual. It implies the whole anatomical and physiological expression of the individual and is not limited to the expression of single characteristics.

The four levels of human biological adaptation take place on different timescales. Acclimatization is the quickest process. It ranges from the immediate regulation of the rate of heartbeat in a short-distance runner to the increase in the haemoglobin concentration over several months or years in people moving to a higher altitude. The timescale of plasticity is the total life-cycle of the individual human being. Plasticity changes are usually induced early in life but they may also be active much later; for example, permanent immunity to an infectious agent can be acquired late in life. The modification of the population structure acts on an intergenerational scale. Whereas reproductive isolation might break down and a panmictic population can be formed in two generations, it usually takes many generations to increase the level of homozygosity appreciably. The change of the gene frequencies in large populations or in whole species is a relatively slow process, usually taking hundreds or thousands of years.

Biological processes take place on all timescales. All four processes of adaptation are involved in the evolution (i.e. the irreversible change to the spectrum of phenotypes over time) of our species. Our ability to glean information about the different levels of the adaptive process is tightly connected to the timescale of the available data. Very rarely, if ever, do the same data contain information on all four aspects of adaptation.

This chapter is concerned with data about the central part of the adaptational time spectrum. We concentrate on processes that do leave marks on the permanent, adult phenotype of the individual; but we are unable to elucidate processes that are so slow that they do not leave clear marks on phenotypical differences within a few centuries. Consequently, we look for the effects of population structure and plasticity. This focus is quite general in human biological studies. It is primarily induced by the sort of data available (in the present case, height of adult males) and by the spatiotemporal frame of the analysis (here, Denmark after AD 1000).

The purpose of this chapter is to develop and illustrate a method to disentangle the effects of two different types of adaptation (namely plasticity and population structure) on the evolution of the human phenotype over several generations. There are two primary sources of inspiration in this venture: the classical development of the theory of quantitative genetics founded by R. A. Fisher (Fisher, 1918) and G. W. Lasker's analysis of the levels of the adaptive process (Lasker, 1969). However, it has not been possible to follow any of these classical approaches strictly. The relationships of interest in the present context are historical and not direct family ties. This means that we will have to move away from the correlation–regression analysis approach of Fisher (1918) to a strategy based on analysis of variance (ANOVA). Further, it has been necessary to split Lasker's concept of natural selection of genotypes into two separate, albeit mutually related, levels of adaptational changes. With these two provisoes the intellectual venture of this chapter has been outlined by Fisher and Lasker.

The theoretical framework
Correlations between different categories of relatives can readily be recalculated to estimates of heritability that have precise interpretations in stable populations (Fisher, 1918; Falconer, 1960). The pedigree relationships between the involved categories of relatives facilitate the direct calculation of the degree and type of genetic relationship. Relations like these do exist between more distantly related ('unrelated') members of separate communities of the same population. However, they cannot be specified. In classical quantitative genetics the specified amounts of shared genetic information act as invariates in the estimation procedures. In the study of similarities and differences among communities it is necessary to define

other invariates (namely, conditions of living and population structure). These invariates characterize the relationship among communities on the group level.

The process that we are going to study occurred during a short episode of human evolution. It took place in a national population that contributed considerably more to the gene pool of surrounding nations that it received genes from such surrounding populations. This means that we can assume constancy of the national gene frequencies of the Danish population from AD 1000 to the 1960s. This fundamental unity of the historical Danish population is not a condition for the analysis, but it does facilitate it. The same type of process – the secular change of the population mean height – has been going on in many other populations. The analysis is carried out conditional on the essentially invariate gene frequencies in the national Danish population over the past 1000 years. It is probable that our conclusions are not critically dependent on this condition and that small changes in gene frequencies from immigration and random genetic drift as a whole are of little consequence.

The main analysis is carried out as an analysis of a natural experiment. The purpose of an experiment is to establish circumstances that permit the collection of information useful for the description of a specified relationship even outside the particular setting of the experiment. Most experiments with specific bearing on human population genetics and plasticity would be highly unethical. It is not justifiable to conduct experiments that profoundly affect the life of many people over several generations. The natural experiment facilitates the analysis and helps to avoid unethical experiments on human populations. A natural experiment cannot be predesigned, so in that sense it is an ordinary observational study. It is a strictly *post hoc* venture. The trick of changing an observational study into a natural experiment has at least partly been developed in epidemiology. It consists of finding invariates in the historical process and measuring the effect of other less constant variables conditioned on the invariates. Many, perhaps most, historical biological processes consist of simultaneous changes in several sets of variables. The historical process that changed the mostly rural population of short people of preindustrial Denmark to the mostly urban recent population of much taller people forms the frame of our natural experiment.

The experiment consists of isolating and estimating the effect of two types of adaptational processes of importance for the phenotype of the Danish population. The development of the structure of the Danish population and its conditions of living and the associated change of the typical phenotype is the function of at least two separate processes: (1) *the improvement of the conditions of living*, leading to the greater realization of the growth potential of the individuals; and (2) *the breaking down of the traditional pattern of spouse selection*, changing the frequencies of different genotypes without altering the basic national gene frequencies.

The term *permissive* will be used for the present conditions of living, which permit a great realization of the individual growth potential, and the term *restrictive* for the environments that do not permit this to the same high degree. This distinction will be treated as if it reflected a dichotomous variable and not points on a continuum. Further, the term *inbred* is used to describe a population structure with great intercommunity genetic variance, most often due to random genetic drift, as its antonym, the term *outbred* is used. The term inbred is used in spite of the fact that considerable conscious effort was put into avoiding consanguineous marriages in preindustrial Denmark (Boldsen, 1989).

The breaking down of reproductive isolation among local communities is expected to have an immediate effect on mean stature. The change in mean stature of Danish men from the late nineteenth century to the present is gradual increase (see Fig. 4.1). This observation appears to contradict the hypothesis about the importance of the population structure effect on mean stature even before it has been estimated. However, it is possible that environmental interference distributed the population structure effect over a few generations. A maternal effect could provide just such an environmental interference.

There is some evidence for a nonlinear maternal effect on height. Boldsen & Mascie-Taylor (1990) showed this to be so in a large sample of British females. When the difference between maternal and paternal height is large the height of the offspring is determined more by maternal than by paternal height. Short women with tall husbands are expected to have offspring that grow much taller than themselves, and tall women with short husbands are expected to have offspring much shorter than themselves. This means that the maternal effect might only indirectly be a consequence of the difference between maternal and paternal height as Boldsen & Mascie-Taylor (1990) described it. Let M denote the height of the mother, F the height of the father, and O the adult height of the offspring. In the arguments given below it is assumed that M, F and O are corrected for sex differences in mean and standard deviation for height. Further, let h^2 denote the heritability for height in the community under study. In a historical period with no intergenerational difference of mean height, the height of the offspring, O can be found as follows:

$$O = [h^2(M + F)/2] + \varepsilon \qquad (4.1)$$

where ε is the individual error term, which has mean zero. The maternal effect described by Boldsen & Mascie-Taylor (1990) can be given the following mathematical interpretation:

$$|O - M| < C_1 \qquad (4.2)$$

where C_1 is constant upper limit to the absolute value of the difference between O and M, the height of the offspring and its mother. Other interpretations of the maternal effect are possible but they will not be discussed here. Now assume that a change in the population structure adds a second constant to the height of the offspring:

$$O' = O + C_2 \qquad (4.3)$$

Substituting O in (4.2) with O' from (4.3) we get:

$$|O + C_2 - M| < C_1 \qquad (4.4)$$

Formula (4.4) described the maternal effect under a secular trend. If C_2 is large relative to C_1, then this maternal effect will reduce the mean difference between mean stature in the parental and the offspring generation.

In a historical phase with rapid changes in the population structure and conditions of living, the maternal effect might modify the population mean height considerably. This means that it may be impossible to detect the immediate effect of socioeconomic or population structural changes on the mean height of the population. It is possible that the introduction of the maternal effect into the interpretation of the patterns of height variation will reconcile the results of quantitative genetic analyses of height (following the line laid out by Fisher, 1918) with those of historical studies of height (e.g. Kill, 1939; Komlos, 1989). However, the maternal effect is only short-lasting. In a few generations it ceases to be of importance. This means that it might interfere with the correlations between first-degree relatives but it is unlikely to have any effect on the population mean height in historical periods when the population structure and the conditions of living remain constant.

Material and methods

There are two main sources for measured male stature in Danish history: stature measurements taken in the graves during excavation (*skeletal heights*) and stature measured at the medical examination prior to conscription (*conscription heights*).

Skeletal heights have been drawn from the excavation of two sites in the Danish main province of Jutland: Sct. Mikkel and Tirup. The Sct. Mikkel site was an urban cemetery in use from the first half of the twelfth century

AD (and perhaps even earlier) until the church on the site was demolished in 1529 (Boldsen, 1979). It was situated just south of the city wall of the mediaeval ecclesiastical centre of Viborg in central Jutland. Most of the graves yielding information useful in the present context were probably laid out in the fifteenth century, as most older graves had been disturbed by later burial activity. Tirup is the only totally excavated rural Danish mediaeval cemetery (Kieffer-Olsen, Boldsen & Pentz, 1986). The village to which the church and cemetery belonged has disappeared completely; it is even unknown from written sources. Based on church architecture and burial custom the use of the cemetery has been dated from around AD 1100 to the first decades of the fourteenth century.

All the skeletal heights have been measured by the author in accordance with a method developed for that purpose during the excavation of the Sct. Mikkel cemetery (Boldsen, 1984a). Large bodies of data on measured femoral lengths exist (Boldsen, 1990b). In spite of the fact that regression formulae for the prediction of living stature from long-bone lengths (Trotter & Gleser, 1952, 1958) are used frequently in the study of past populations in Denmark (e.g. Bennike, 1985) it has been shown that these formulae miss the target height by 3.2 cm among males and 1.4 cm among females from Sct. Mikkel (Boldsen, 1984a). Further, it has recently been shown that the formulae developed on the basis of the graves and skeletons from the Sct. Mikkel excavation do not fit the simultaneous distribution of femoral length and stature in the Tirup sample (Boldsen, 1990a). In fact, it appears to be a general problem that regression of stature on long-bone lengths produces non-normally distributed residuals in Danish mediaeval samples. This last observation invalidates the regression approach to the prediction of living stature from long bones. Therefore the only skeletal stature measurements discussed in this chapter are those based on direct measurements in the grave. Estimates for other cemetery sites based on long-bone dimensions alone are not included in this discussion.

It is possible to obtain fairly accurate and reliable information on mean male stature back to the beginning of the ninetennth century in Denmark. At that time only rural men were conscripted; but they formed some 80% of the total population (Johansen, 1991). There is evidence for a long-lasting difference between the mean stature of rural and urban Danish men (Boldsen, 1990b; Westergaard, 1911). This means that it is impossible accurately to interpret the national mean heights of the men measured in 1815, 1825, and 1835 (Thune, 1848). Table 4.1 summarizes most of what is known about the stature of Danish men from the Middle Ages to the present. Only samples assumed to be representative for either the rural, the urban or the whole national population have been included. This means

Table 4.1. *Mean measured male heights from different parts of Danish history*

Year of birth (approximate)	Mean measured male heights (cm)		
	urban	rural	national
1100–1500	172.2	166.2	166.8 (estimated)
1790–1797	—	165.3	—
1800–1807	—	167.0	—
1810–1817	—	166.5	—
1820–1827	—	166.5	—
1827–1838	—	—	166.3
1877–1892	169.9	169.1	169.4
1905	—	—	169.5
1915	—	—	170.8
1925	—	—	173.2
1935	—	—	173.8
1945	—	—	175.9
1955	—	—	178.3
1965	—	—	179.4

These data illustrate the problem (an increase of more than two standard deviations in mean male height in Denmark in only 140 years).

that samples of measured heights of soldiers have been excluded. For the men born from 1790 to 1965 all measurements were taken during the compulsory examination prior to conscription into the armed forces. The source for the men born from 1790 to 1827 is Thune (1848). The information on men born from 1827 to 1838 originates from *Meddelelser fra det Statistiske Bureau* (1859). The data on men born after 1877 originates from material in *Danmarks Statistik*. Fig. 4.1 illustrates the state of knowledge about mean male height in historical Danish populations.

Sample classification
The samples are classified both in accord with the conditions of living and growing and in accord with the population structure in the historical populations that they represent. In this classification and in the rest of this paper we shall concentrate on the study of samples that are thought to represent fairly stable conditions. The present-day sample of the national Danish population is included as one of the samples representing stable conditions. Of course, it is not known, yet, whether this classification is correct or if the future will see more dramatic fluctuations in the conditions of living and growing of the Danes; but they are our best estimate of the Danish population at the end of the secular trend for mean male height.

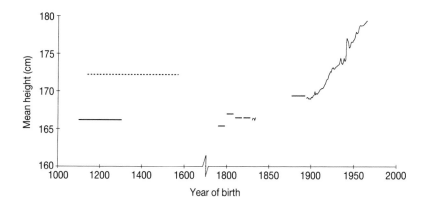

Fig. 4.1. Mean male height in historical Danish samples. The broken line is the mediaeval urban sample; all other samples are either rural or national.

The conditions of living probably did not remain constant throughout the time when the two mediaeval cemeteries were in use. However, the skeletons coming from those sites were sampled over at least two centuries so it is assumed that they represent stable or at least average conditions of living and growing. These two samples are also included in the analyses below, as are the samples of measured heights of rural men born between 1790 and 1827. It appears that these samples represent the Danish population at the starting point of the secular trend.

Conditions of living and growing

Growth is very difficult to study in past populations. It is impossible to follow any individuals longitudinally. Cross-sectional data are available; but they are of a very peculiar sort as the cross-section is at death. However, there is some evidence that most childhood mortality was due to acute conditions that had not interfered with the normal rate of growth (J. L. Boldsen, J. Kieffer-Olsen & P. Pentz, work in progress). It is generally difficult or impossible to determine the sex of an immature skeleton. However, growth to the age of 10 years is reasonably parallel in boys and girls (see Andersen *et al.*, 1982). Growth between 10 and 20 is dominated by the pubertal growth spurt and the age at peak height velocity (PHV) is usually lower in females than in males. This means that it is impossible to construct growth curves for youths between 10 and 20 years based on skeletal data. Fig. 4.2 shows the reconstructed cross-sectional growth curves from age 1 to 10 years for the two mediaeval skeletal samples and for contemporary Danes. The skeletal heights have been measured in the

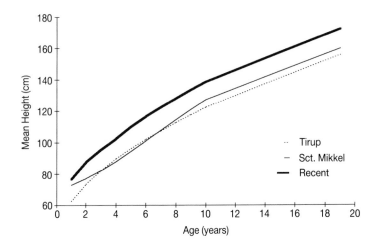

Fig. 4.2. Growth curves for recent Danish children (boys and girls averaged), the Sct. Mikkel children and the Tirup children.

graves in the same way as the adult heights were measured (see Boldsen, 1984a); data on the contemporary Danes have been extracted from Andersen *et al.* (1982). The curve for the contemporary Danes has been drawn from the raw data whereas the curves for the two mediaeval samples are the heights predicted from a polynomial regression of height upon age at death. This procedure has been followed owing to the small number of observations in each age category in the two samples.

The patterns of growth are grossly different in the three samples. The contemporary Danes are tallest at all ages. The children from the Tirup cemetery appear to follow a growth trajectory quite parallel to that of contemporary Danes, constantly being some 14 cm shorter. The pattern of growth is quite different in the Sct. Mikkel sample. It appears that the urban children in this sample grew slowly during their second to sixth years and faster thereafter. In the middle of the nineteenth century, Honer collected cross-sectional data on the height of children working in factories in United Kingdom (see Tanner, 1981). The youngest children included in this survey were 8 years old. This means that there is only limited overlap between the Honer data and the data presented in Fig. 4.2. However, it appears that growth pattern of the nineteenth century children experiencing the hardship of factory work closely follows the growth patterns of the children from the urban cemetery of Sct. Mikkel.

It looks as if the rural children from Tirup were born small and grew 'normally' whereas the urban children were born with more or less modern sizes but experienced postnatal growth retardation concentrated in early childhood. At late childhood the Tirup children were 3–4 cm shorter than the British children in the Honer sample and the urban children from Sct. Mikkel. It is interesting to note that the difference of adult stature among the Danish samples was established by the age of 10 years. It appears that the pubertal growth spurt did not play any great part in differentiating the adult height among the three samples of Danes.

The profound and lasting negative effect of the conditions of living and growing is clearly seen in the growth curve for the Sct. Mikkel children. It can be assumed that the Tirup children experienced environmentally induced growth retardation, too. However, this retardation is more likely to have been concentrated in the pre and early postnatal period and thus cannot be demonstrated in Fig. 4.2. Following the analysis of the patterns of growth, it appears that both the Tirup and the Sct. Mikkel populations lived under restrictive conditions. If anything the evidence indicates even worse conditions of living and growing in the urban community, although the people in it end up taller than in the rural community. It is impossible to determine whether the people from Tirup would have been even shorter had their early childhood conditions of living been the same as those the Sct. Mikkel children experienced. It is likely that the restrictive elements of the environment were different in rural and urban communities.

Population structure

Danish measured male heights have been shown to deviate from the normal distribution in a characteristic way. Boldsen & Kronborg (1984) have demonstrated the presence of a short subpopulation of young men born in the first half of the nineteenth century in Denmark. The frequency of this short subpopulation declined from around 0.6% in the earliest samples to 0.1% in samples of men born after 1950 (Boldsen, 1983). A reanalysis of the Boldsen (1983) figures indicates that the height deviates from normality even after the mixture effects have been corrected for in the way that Boldsen & Kronborg (1984) did. It appears that a mixture of log-normal distributions is a more valid model for the Danish conscription heights. This change of model has hardly any influence on any of the distribution parameters, except the variance (σ^2). After correcting for the mixed nature of the height distribution it is possible to correct the estimate for σ^2 of the log-normal distribution by calculating the relative variance, λ, in the following way:

$$\lambda = \log_e(1 + \sigma^2/\mu^2),$$

where μ is the mean height. Fig. 4.3 illustrates the development of the estimated value of λ over time. The value of λ reduces by some 17.5% from

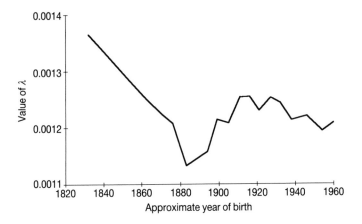

Fig. 4.3. The development of λ (the relative variance for height) from the 1830s to the 1950s.

the first half of the nineteenth century to the 1880s. In Holland, van Wieringen (1986) found that the increase of mean stature in the nineteenth century was associated with a reduction of the range. A reduction of the range for height associated with an increase of mean height corresponds to a decline of the value of the λ-statistic. So the first part of Fig. 4.3 illustrates a development that is not unique to Denmark. After the dramatic decline in the relative variance in Denmark in the latter part of the nineteenth century, it increases once again and stabilizes around half way between the two extreme values in men born in this century. This development is exactly what one would expect to see in a population moving from an inbred stage through a heterotic generation to a stable outbred stage. In fact, the development of the relative variance permits the estimation of the amount of genetic variance that was due to genetic differences between local communities. If the broad-sense heritability for height is close to 100% in each cohort (see Fisher, 1918), then the between-local-community genetic variance component (F_B) can be estimated as follows:

$$F_B = (\lambda_{max} - \lambda_{min})/(2\lambda_{stable}),$$

where λ_{stable} is the mean of the relative variances for men born in this century. Using the values illustrated in Fig. 4.3, F_B is found to be 0.095. The genetic effect of F_B is the same as the effect of inbreeding: it increases the frequency of homozygosity in the population. The difference lies only in the process leading to this result.

The development of λ illustrated in Fig. 4.3 indicates that the Danish population born up to the first half of the nineteenth century was highly inbred, that the generation born between 1870 and 1900 was crossbred (heterotic), and that men born in this century are derived from a (genetically speaking) rather stable outbreeding population.

Simulation studies have been carried out using the Tirup mortality profile as the demographic basis of hypothetical mediaeval village communities and using the canonical rules for the avoidance of consanguineous marriages as the basis for mate selection (Boldsen, 1988, 1989). In the simulations it is assumed that migration will only take place if there is no suitable mate in the community. The results indicate that small communities with stable population sizes of 75 persons like the Tirup community (see Kieffer-Olsen *et al.*, 1986) have high migration frequencies and large communities with stable population sizes of 300 require virtually no migration. The effect of population size on the migration frequency makes the level of inbreeding (measured as F_B) more or less independent of population size. The F_B estimated from the simulation study is 0.049, only around half of the value estimated from the study of variances for height. In the simulations it is assumed that the immigrants were drawn from distant and totally unrelated communities. This is probably not a realistic model for the true but unknown patterns of migration between mediaeval rural communities. If preferred routes of migration existed they would lead to higher levels of F_B than found in the simulations. At any rate, both the simulation study and the study of the development of λ clearly indicate that the rural communities in the Middle Ages and in the first half of the nineteenth century were highly inbred.

It has been shown that the mediaeval city of Viborg (where the Sct. Mikkel cemetery was situated) could not maintain its population without constant immigration of young adult women (Boldsen, 1984b). Further, the population size of this ecclesiastical centre was many times that of the mediaeval villages. This means that the results of the simulation study have no bearing on the genetic population structure of the Sct. Mikkel community. The demographic conditions of this community effectively prevented the development of a F_B of any importance. The conclusion must be that the Tirup sample of skeletal heights and the sample of heights measured on men born in the first half of the nineteenth century were drawn from highly inbred populations whereas the Sct. Mikkel skeletal sample and the recent sample of measured heights were drawn from outbred populations.

Plasticity and population structure effects
It is not possible to fill in all four cells in Table 4.2, as no sample of Danes who grew up under permissive living conditions in an inbred community is

Table 4.2. *Mean heights of Danish males broken down by conditions of living and by population structure*

	Conditions of living and growing	
Population structure	permissive	restrictive
inbred	—	166.3
outbred	179.4	172.2

available. Insufficient sampling is not the reason for the lack of data: the combination has probably never existed in Danish population history. This fact excludes the estimation of a possible interaction effect between conditions of living and population structure. However, it is possible to estimate the effects under the historical process that the Danish population went through. It would be of considerable theoretical interest to be able to estimate the effects if the conditions of living had changed before the population structure changed.

The effect estimates can be derived directly by subtraction of the mean heights given in Table 4.2. The plasticity effect (PL) is estimated by the difference between the mean height of men living under permissive conditions in an outbred population from the mean height of men living under restrictive conditions in an outbred population, $PL = -7.2$ cm (s.e.e. $= 1.9$, $t = -3.77$, d.f. $= 30$, $0.001 > p > 0.0005$). In the same way the population structure effect (PS) can be estimated by subtracting the mean height of the inbred populations living under restrictive conditions from the mean height of the outbred population that also lived under restrictive conditions, $PS = 5.9$ cm (s.e.e. $= 1.9$, $t = 3.08$, d.f. $= 30$, $0.01 > p > 0.005$).

The fourth cell in Table 4.2 cannot (owing to the maternal effect) be filled in by the heights of children born to related couples. Such children would have little or no bearing on the mean height of men coming from an inbred population living under permissive conditions.

The estimates derived in this chapter are only the first approximations of the true values. Much more work needs to be done before solid estimates can be reached. The main purpose of this chapter has been to show that it is possible to partition the secular trend for male stature in one country into a plasticity and a population structure component. The values of the estimates are at present of less concern. If any thing, the 5.9 cm for the population structure effect could be an underestimate and, consequently, the 7.2 cm for the plasticity component would be an overestimate.

Several different types of studies and lines of empirical evidence must be considered in the ongoing research into the reasons for secular changes in the human phenotype. Firstly, more and better classified mediaeval material is needed. Secondly, it is necessary to move beyond the restrictive–permissive dichotomy for the conditions of living and growing and try to reach continuous measures of the quality of the environment. Thirdly, whole population data, such as the conscription heights utilized in this chapter, must be combined with exact econometric data to elucidate the short-term effects of fluctuations in the economic aspects of the environment. Work along these lines is in progress: skeletons are being excavated and measured; the understanding of the mediaeval burial pattern has increased; and time-series analyses are being carried out using the Danish conscription data and economic history. However, new ideas and approaches are needed to reach conclusions that can be accepted by both quantitative geneticists and anthropometric historians.

References

Andersen, E., Hutchins, B., Jansen, J. & Nyholm, N. (1982). Højde og vægt hos danske børn. *Ugeskrift for Læger* **144**, 1760–5.

Bennike, P. (1985) *Palaeopathology of Danish Skeletons. A. Comparative Study of Demography, Disease and Injury*. Copenhagen: Akademisk Forlag.

Boldsen, J. L. (1979). Liv og død i middelalderens Viborg. *Museer I Viborg Amt* **8**, 76–85.

Boldsen, J. L. (1983). Studier af humanbiologisk variation i fortid og nutid – metoder og resultater. Dissertation, Department of Theoretical Statistics, University of Aarhus.

Boldsen, J. L. (1984a). A statistical evaluation of the basis for predicting stature from lengths of long bones in European populations. *American Journal of Physical Anthropology* **65**, 305–11.

Boldsen, J. L. (1984b). Palaeodemography of two southern Scandinavian Medieval communities. *Meddelanden från Lunds universitets historiska museum 1983-1984* (new series) **5**, 105–15.

Boldsen, J. L. (1988). Two methods for the reconstruction of the empirical mortality profile. *Human Evolution* **3**, 335–42.

Boldsen, J. L. (1989). Vejen til byen – En skitse af de befolkningsmæssige relationer mellem land og by i det middelalderlige Jylland belyst ved skeletfund. *Land og By i Middelalderen* **5–6**, 127–59.

Boldsen, J. L. (1990a). Population structure, body proportions and height prediction. *Journal of Forensic Medicine* **6**, 157–65.

Boldsen, J. L. (1990b). Height variation in the light of social and regional differences in Medieval Denmark. In: *From the Baltic to the Balck Sea: Studies in Medieval Archaeology*, ed. L. Alcock & D. Austin, pp. 181–8. London: Unwin Hyman.

Boldsen, J. L. & Kronborg, D. (1984). The distribution of stature among Danish conscripts 1852-1856. *Annals of Human Biology* **11**, 555–65.

Boldsen, J. L. & Mascie-Taylor, C. G. N. (1990). Evidence for maternal inheritance of female height in a national British sample. *Human Biology* **62**, 767–72.

Falconer, D. S. (1960). *Introduction to Quantitative Genetics*. Edinburgh: Oliver & Boyd.

Fisher, R. A. (1918). The correlation between relatives on the supposition of Mendelian inheritance. *Transactions of the Royal Society of Edinburgh* **52**, 399–433.

Johansen, H. C. (1991). *Danmark i tal. Gyldendal og Politikens Danmarks Historie*, vol. 16. Copenhagen: Gyldendal Politiken.

Kieffer-Olsen, J., Boldsen, J. L. & Pentz, P. (1986). En nyfunden kirke ved Bygholm. *Vejle Amts Årborg* 1986, 24–51.

Kill, V. (1939). *Stature and Growth of Norwegian Men During the Past Two Hundred Years. (Skrifter utgitt av Det Norske Videnskaps-Akademi i Oslo, I, Mat. -Nat. Klasse* 1939, no. 6.)

Komlos, J. (1989). *Nutrition and Economic Development in the Eighteenth Century Habsburg Monarch – An Anthropometric History*. Princeton, New Jersey: Princeton University Press.

Lasker, G. W. (1969). Human biological adaptibility. The ecological approach in physical anthropology. *Science* **166**, 1480.

Mascie-Taylor, C. G. N. & Boldsen, J. L. (1985). Regional and social analysis of height variation in a contemporary British sample. *Annals of Human Biology* **14**, 59–68.

Meddelelser fra det Statistiske Bureau 1859 (1859), Femte samling, pp. 159–90.

Tanner, J. M. (1981). *A History of the Study of Human Growth*. Cambridge University Press.

Thune, L. G. W. (1848). Den fuldvoksne værnepligtige bondeungdoms legems højde i Danmark. *Det kongelige medicinske Selskabs Skrifter. Ny Række* vol. 1.

Trotter, M. & Gleser, G. C. (1952). Estimation of stature from long bones of American whites and Negroes. *American Journal of Physical Anthropology* **10**, 463–514.

Trotter, M. & Gleser, G. C. (1958). A re-evaluation of estimation of stature based on measurements of stature taken during life and long bones after death. *American Journal of Physical Anthropology* **16**, 79–123.

Weiringen, J. C. van (1986). Secular growth changes. In: *Human Growth–A Comprehensive Treatise* (2nd edn), vol. 3, ed. F. Falkner & J. M. Tanner, pp. 307–31. New York: Plenum Press.

Westergaard, H. (1911). Undersøgelser over legemshøjden i Danmark. *Meddelelser om Danmarks Antropologie* **1**, 351–402.

5 Plasticity, growth and energy balance

STANLEY J. ULIJASZEK

Summary
There is huge variation in human energy expenditure world-wide; in this chapter, an attempt is made to see whether any of it can be attributed to plasticity. Most of the variation is due to differences in body size and physical activity. The former is self-evident, and can clearly be seen as an outcome of growth and development, while the second is largely behavioural, and is not subject to plasticity. More subtle factors, including variation in body composition, increased efficiency of muscular activity and increased metabolic efficiency at the cellular level, are the subjects of considerable research and debate, and are considered here in relation to the hormonal mechanisms that operate in growth faltering. The lower than expected basal metabolic rate in some populations cannot be attributed to body composition changes (i.e. plasticity) which arise in the course of growth faltering. However, economization of energy expenditure through increased muscular efficiency could arise during growth faltering. Low basal metabolic rate due to increased metabolic efficiency at the cellular level could be an indirect outcome of the lower food intake required to maintain energy balance at lower body weight. Both are likely to be mediated by thyroid hormones, which have a wide field of operation, and are directly involved in the growth faltering process itself.

Introduction
There is considerable variation in human energy expenditure, within and between individuals, and between populations. One accommodation to low energy intake in adults is small body size (Blaxter & Waterlow, 1985). If plasticity is the irreversible development of a range of characteristics in the course of growth and development, it is reasonable to question which nutritional adaptations, if any, could be outcomes of this process.

Low energy intake
Typical Western and Third World growth patterns are known to differ, inasmuch as linear growth retardation often occurs in the latter (Waterlow, 1988). For children in developing countries, slow growth is associated with

91

greater risk of death and disease (Pelletier, 1991; van Lerberghe, 1989).

In childhood, low availability of dietary energy is often associated with growth faltering, which if prolonged will result in smaller overall adult body size. Physiological adaptations (not necessarily cost-free) in childhood include a reduction in growth rate, metabolic rate and protein turnover, and are mediated by insulin-like growth factor I (IGF-I), and the principal thyroid hormone, triiodothyronine (T3).

Energy and protein intakes are important factors in the regulation of serum IGF-I (Isley, Underwood & Clemmons, 1983; Underwood et al., 1989). IGF-I has bone as a primary target, and this is perhaps the most sensitive pathway by which human growth may be restricted by low food intake. Certainly, during fasting, serum IGF-I may drop to 40% of usual, non-fasting levels by 5 days.

The fall in T3 in response to low food intake occurs within 24 h, reaching values 40% lower than in normally fed subjects within 48–72 h (Danforth, 1985). At the cellular level, the action of T3 may involve two interrelated processes. The first is its effect on metabolic rate and consequent energy balance. The second is on growth and development, through its influence on gene expression, protein synthesis, and protein turnover (Millward, 1986). In the pituitary, thyroid hormone regulates the synthesis of growth hormone (Danforth & Burger, 1989). Although a fall in T3 concentration has been linked to an observed decline in resting metabolism under experimental conditions (Keys et al., 1950), the sympathetic nervous system (SNS) may be more important in regulating changes in resting metabolism under conditions of positive and negative energy balance (Danforth & Burger, 1989).

Undernourished children can conserve nitrogen from protein when their energy intakes fall (Jackson, 1985) and rates of whole-body protein turnover are depressed (Golden, Waterlow & Picou, 1977), suggesting that reduced protein degradation, as a consequence of lower T3 activity or concentration, may be a response to undernutrition. However, during growth a fall in T3 has a greater inhibitory effect on protein synthesis than on degradation, inhibiting growth rather than sparing it (Millward, 1986). Thus, there is a balance between possible homonally mediated metabolic efficiencies and the growth faltering process.

Components of energy expenditure

Human energy expenditure can be broken down into a number of components, including the energy cost of maintenance, thermic effect of food, physical activity, thermoregulation, growth, and reproduction. These components vary between individuals and populations, and differ according to the extent to which they can be modified by human behaviour.

The energy cost of bodily maintenance includes all functions that preserve bodily integrity, including cardiac activity, sodium and calcium pumping, sympathetic nervous system activity, and homeostatic metabolic processes, including thermoregulation (Ulijaszek, 1992). An approximation of this can be obtained by the measurement of basal metabolic rate (BMR). The energy cost of digestion, absorption, transport, metabolism and storage of ingested food is called the thermic effect of food (TEF), and accounts for approximately 10% of energy intake (Horton, 1983). It can only be estimated under controlled laboratory conditions and is rarely measured in the field, although it has been shown that measures similar to TEF, dietary-induced thermogenesis and post-prandial thermogenesis, vary between 6 and 8%, and 5 and 6% of energy intake across seasons for Gambian men (Minghelli *et al.*, 1991) and women (Frigerio *et al.*, 1992), respectively. The TEF is influenced by the macronutrient composition of the diet and by the level of energy intake (Belko, Barbieri & Wong, 1986; Kinabo & Durnin, 1990; Norgan, 1990). Energy is expended in physical activity by all individuals at all times, while growth and reproduction carry energetic costs in childhood, and in most adult women at some time, respectively.

Economization of energy expenditure

There are three possible ways in which energy expenditure can be economized: (1) low body weight; (2) reduction in the amount, or the cost, of physical activity; and (3) increased biochemical efficiency (Waterlow, 1990). In addition, there may be mental states, such as apathy due to dietary restriction, which are appropriate to conserving energy. The extent to which each of the above can contribute to such economies can be illustrated by examining data from the Minnesota starvation experiment of 1944–5 (Keys *et al.*, 1950).

In the Minnesota study, 32 adult male consciencious objectors were partly starved, on a dietary intake of half their habitual levels, for 24 weeks. Fig. 5.1 shows mean body size and composition at 0, 12 and 24 weeks of study; Fig. 5.2 gives mean energy balance and expenditure during the same time. By 12 weeks of semi-starvation, mean weight was 83% of its initial value, while by 24 weeks, the value was 75%. The proportion of body weight as fat was 14% initially, declining to 8% and 6% at 12 and 24 weeks of semi-starvation, respectively. The 12 and 24 week values were 46% and 31% of the value at the start of the study. Fat-free mass (FFM) was 88% and 83% of the initial value at 12 and 24 weeks, respectively. Thus the decline in body weight came predominantly, but not totally, from a loss of body fat.

There was large negative energy balance at the start of the study, in excess

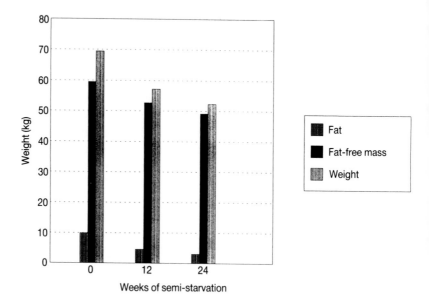

Fig. 5.1. Minnesota starvation experiment: mean body weight, fat, and fat-free mass at 0, 12, and 24 weeks of semi-starvation.

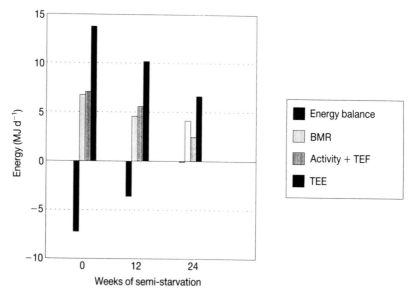

Fig. 5.2. Minnesota starvation experiment: mean energy balance and energy expenditure at 0, 12, and 24 weeks of semi-starvation (BMR, basal metabolic rate; TEF, thermic effect of food; TEE, total daily energy expenditure).

Table 5.1. *Basal metabolic rate of subjects taking part in the Minnesota starvation experiment*

	BMR (MJ d^{-1}) mean (SD)	BMR kg^{-1} (MJ d^{-1}) mean (SD)	BMR kg^{-1} FFM (MJ d^{-1}) mean (SD)
Week of study:			
0	6.87 (0.41)	97.2 (7.3)	113.1 (7.0)
12	4.62 (0.45)	80.5 (7.0)	87.7 (10.3)
24	4.19 (0.42)	79.8 (8.1)	84.8 (9.7)
One-way ANOVA: *F* ratio		55.5*[a]	93.4*
Scheffé test,			
week 0 vs. week 12		*	*
week 12 vs. week 24		NS[a]	NS

[a] $*p < 0.001$; NS, not significant.
Source: Keys *et al.* (1950).

of 7 MJ per day (about 1700 kcal), which was reduced to a very small value at 24 weeks. Thus the subjects did not, on average, achieve energy balance at their lower level of intake. Total energy expenditure fell from about 15 MJ d^{-1} (3400 kcal d^{-1}) to about 7 MJ d^{-1} across the 24 weeks. Most of this was due to loss of body mass, and of metabolically active tissue in particular.

During the first 12 week period, there was a decline in BMR, energy expended in physical activity (ACT) and TEF, all occuring in more-or-less equal proportion to total energy expenditure (TEE). Between weeks 12 and 24, a further decline in TEE came from a greater decline in ACT relative to BMR, the decline in TEF being in about the same proportion to TEE as in weeks 0 to 12. The fall in BMR across this period was much smaller than across the earlier period. At the end of 24 weeks' partial starvation, energy expended in physical activity as a proportion of total energy expenditure was lower than at 0 or 12 weeks.

Table 1 gives mean BMR per kilogram of body weight, and per kilogram of FFM, across the period of study. Both fell significantly between weeks 0 and 12, but not between weeks 12 and 24. The 24 week value for BMR per kilogram body weight was 18% lower than the value at week 0, while for BMR per kilogram FFM the 24 week value was 25% lower than that at the start of the study. Thus there appeared to be considerable down-regulation

Table 5.2. *Mean energy cost of lying, sitting, and standing, and of stepping at different rates* $(kJ\ min^{-1})$

Activity (n)	Gurkha	British	Paired *t*-test[a]
Lying (11)	5.6	5.5	—
Sitting (11)	6.3	5.9	—
Standing (11)	6.7	6.8	—
15 steps min^{-1} (15)	27.1	36.2	**
20 steps min^{-1} (15)	37.6	46.4	**
25 steps min^{-1} (13)	51.2	56.4	*
30 steps min^{-1} (10)	50.8	58.5	—

[a] Significance levels: *, $p < 0.05$; **, $p < 0.01$.
Source: Strickland & Ulijaszek (1990).

Table 5.3 *The percentage by which Schofield's (1985) equations overestimate basal metabolic rate in a number of populations, ages* 18–60 years

Country or nationality	Males		Females	
	sample size	mean (%)	sample size	mean (%)
Sri Lanka	125	22.4	100	12.5
India	50	12.8	—	
Philippines	172	9.5	—	—
South America	941	9.4	227	4.8
Malaysia	62	9.3	—	—
Hawaii, non-European	—	—	62	4.5
Japanese	202	5.8	152	4.6
Javanese	86	5.0	—	—
Mayan	76	1.5	—	—

Source: Henry & Rees (1991).

of BMR, perhaps as a result of increased metabolic efficiency at the cellular level.

This study shows the three ways in which energy economies can be made in response to a decline in food availability in adults. Other studies have shown energy economies in populations which habitually experience low energy intakes. These include the following: (1) lower than expected energetic costs of physical activity (Ashworth, 1968; Maloiy *et al.*, 1986; Jones *et al.*, 1987; Strickland & Ulijaszek, 1990; Kulkarni & Shetty, 1992); (2) lower than expected BMR (Henry & Rees, 1988; 1991); and (3) smaller body size relative to well-nourished European populations (Eveleth & Tanner, 1990).

In a study of work performance in Gurkha and British soldiers, Strickland & Ulijaszek (1990) found that the Nepali soldiers, matched with the British ones by body weight, expended less energy in performing the same step-test, but not in lying, sitting, and standing (Table 5.2). They speculated that variations in thyroid hormone status during growth and development could have influenced the muscular efficiency of the Gurkhas, by its effect on the skeletal muscle phenotype.

Henry & Rees (1988, 1991) found that regression equations obtained from meta-analysis of BMR and body weight from a large number of studies carried out predominantly on European subjects (Schofield, 1985) tended to overpredict mean BMRs of tropical populations by up to 22% (Table 5.3). They discounted the possibility that the lower BMR observed in various groups could be due to the inclusion of malnourished subjects. However, the mean weights of these tropical groups were considerably lower than those of European populations, and the possibility that undernutrition in the course of growth and development could have influenced BMR cannot be discounted.

Smaller size means lower energy cost of bodily maintenance and physical activity involving the movement of the body. Most of the world's populations have mean body weights lower than those of European and other well-off populations, largely because of the growth-faltering effects of undernutrition and infection in childhood. That is, small body size arises from the plastic response to those stresses. Is it possible that the lower energetic costs of physical activity and BMR already described could be due to plasticity?

Growth faltering and muscular development

Although the metabolic rate of skeletal muscle at rest is low in comparison with organs such as the brain, kidneys and liver (Table 5.4), its large mass in adults means that it makes a significant contribution to oxidative metabolism. Small improvements in metabolic efficiency may therefore make important contributions to energy economy in an individual experiencing low food intake. Muscles with a high proportion of slow-twitch fibres use less ATP per unit of isometric tension than muscles containing a high proportion of fast-twitch fibres (Wendt & Gibbs, 1973; Goldspink; 1975), although during dynamic contractions, the energy cost to a fibre type also depends on the speed of contraction (Nwoye & Goldspink, 1981). Following from this, atrophy of fast-twitch fibre mass might be energetically advantageous, provided that rapid movements are avoided (Henriksson, 1990).

Table 5.4. *Metabolic rates of different bodily tissues*

Tissue	kJ kg^{-1} d^{-1}
Heart	1840
Kidneys	1840
Brain	1000
Liver	840
Muscle	50
Adipose	20
Miscellaneous[a]	50

[a] Sum of bone, skin, intestines, glands, by difference.
Source: Elia (1992a).

Studies of low-intensity, long-duration training in healthy adults showed a reduction in fast-twitch fibres in the triceps brachii and quadriceps muscles, whereas slow-twitch fibres were not affected (Schantz, Henriksson & Jansson, 1983). Similarly, investigations of muscle size and composition in malnourished patients showed that the size of the slow-twitch fibres in the calf muscle was better preserved than that of the fast-twitch fibres. In the developing world, low dietary energy availability and prolonged low-intensity work output may lead to such adaptations in muscular development, and the economization of energy expenditure made possible by them.

Differential development of slow-twitch over fast-twitch fibres in the course of growth faltering has not been demonstrated. However, muscle phenotype characteristics may be modified or determined by thyroid hormone status (Nicol & Bruce, 1981; Jolesz & Sreter, 1981). Notably, clinically hypothyroid patients show significantly lower energy expenditure compared with euthyroid controls during quadriceps muscle function tests (Wiles *et al.*, 1979). A possible mechanism for this effect might be that thyroid hormones influence the sequestration and release of calcium ions by the sarcoplasmic reticulum, and ATP consumption in calcium pumping during the contraction cycle, thereby increasing overall efficiency (Suko, 1973; Limas, 1978; van Hardeveld & Clausen, 1984). It is not clear whether the disappearance of the low thyroid status associated with growth faltering leads to reversion of the muscle phenotype to one with a greater proportion of fast-twitch fibres. If nutritional stress persists through life, then low T3 will persist, and with it, greater muscular efficiency associated with selective persistence of slow-twitch fibres over fast-twitch fibres. Work is needed to test the proposition that greater economy of muscular activity through differential fibre growth and preservation is, at least in part, a function of plasticity.

Plasticity and basal metabolic rate

The BMR forms a large part of TEE, mean group values ranging between 43% (Viteri *et al.*, 1971) and 73% (Prentice *et al.*, 1985) of TEE. Small improvements in cellular metabolic efficiency are likely therefore to make significant savings in TEE. Interpopulation variation in BMR is well documented (Henry & Rees, 1988; Quenouille *et al.*, 1951), and mean BMRs below values predicted from body weight have been reported (Table 5.3). Although variation in BMR becomes smaller when body weight is taken into consideration, values lower than expected still exist; a variety of reasons for the remaining variation have been suggested. These include body composition, diet, environmental temperature and hormonal differences (Elia, 1992). However, all these factors, apart from body composition, are related in that thyroid hormones have a central role in cellular metabolic adaptation and are influenced by dietary quantity and quality and environmental temperature. It is impossible to determine the effects of small differences in body composition on BMR with the techniques and technologies currently available, since it would be necessary to measure the size of the highly metabolically active tissues of the liver, brain and kidneys, as well as the body compartments it is now possible to measure with some degree of success (Shephard, 1991).

To examine the possibility that BMR differences between larger and smaller subjects may be outcomes of developmental plasticity, it would be necessary to accurately measure the body composition of children who show no growth faltering and children who do, across their developmental span and into adult life. Current techniques do not allow this; however, it is possible to obtain some approximations by using modelling techniques.

Modelling the effects of body composition changes arising from growth faltering in early childhood, on basal metabolic rate

The modelling procedure exmployed here involves making a number of assumptions, and the results are speculative. However, in the absence of data, this is one way of examining the possibility that the lower BMRs found in adults in the developing world may be a function of body composition differences arising in the course of growth and development, and therefore a result of plasticity.

An estimate of body composition, as a proportion of total body weight at different ages, is given for an individual following the fiftieth centile of weight for age (National Center for Health Statistics, 1977) in Table 5.5. This shows that, in the process of growth and development, there is a doubling in the proportion of total body weight that is skeletal muscle between birth and maturity, with the proportion of extracellular fluid being

Table 5.5. *Estimated body composition of an ideal male, following the National Center for Health Statistics fiftieth centile from birth to 18 years of age*

Data are tissue weight, as a percentage of total body weight.

Age (years)	Weight (kg)	Organs[a]	Skeletal muscle[b]	Body fat	ECF	Everything else[c]
0	3.3	18	20	12	40	10
0.25	6.0	15	22	11	32	20
0.5	7.8	14	23	20	26	17
1	10.2	14	23	20	26	17
2	12.6	13	27	18	25	17
5	18.7	10	35	15	24	16
10	31.4	8	37	15	25	15
18	68.9	5	40	15	19	21

[a]Sum of brain, liver, heart and kidney.
[b]Derived from creatinine excretion.
[c]Bone, cartilage, viscera, by subtraction.
Source: After Holliday, (1986).

Table 5.6. *Estimated body composition of an individual who begins life with weight on the fiftieth centile of NCHS, and who experiences growth faltering in infancy*

Data are tissue weight, as a percentage of total body weight.

Age (years)	Weight (kg)	Organs[a]	Skeletal muscle	Body fat	ECF	Everything else[b]
0	3.3	18	20	12	40	10
0.25	6.0	15	22	11	32	20
0.5	6.9	15	25	10	29	21
1	8.8	16	31	10	25	17
2	11.0	15	33	10	25	17
5	15.9	12	38	10	24	16
10	25.4	10	40	10	25	15
18	57.0	6	44	10	199	21

[a]Sum of brain, liver, heart and kidney.
[b]Bone, cartilage, viscera, by subtraction.
Source: After Holliday, (1986).

reduced by half. Body fatness peaks between the ages of six months and one year, while organ (brain, liver, heart, kidneys) weight as a proportion of total body weight shows a steady decline across childhood and adolescence.

Table 5.6 gives an estimate of body composition of a male child who

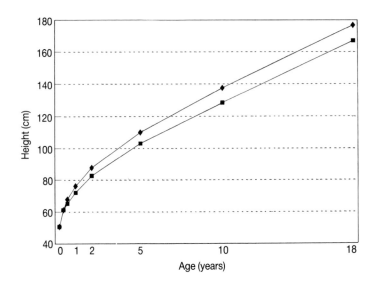

Fig. 5.3. Heights of the two children used in the model. One (diamonds) follows the fiftieth centile of height for age, National Center for Health Statistics (1977); the other (squares) shows growth faltering in late infancy and does not recover the deficit.

begins life on the fiftieth centile of the NCHS references of weight for age, and who experiences growth faltering in infancy, resulting in reduced body size throughout subsequent life (Figs. 5.3 and 5.4). The assumptions made are as follows.

1. That extracellular fluid, as a proportion of total body weight, is marginally higher than normal during the period of growth faltering, but subsequently reverts to that which is expected for the rest of childhood. This takes account of what is known about changes in extracellular fluid content in children with moderate malnutrition, and their subsequent recovery from it (Waterlow, 1992).

2. That body fatness does not peak in late infancy. Rather, body fatness is assumed to decline to 10% of total body weight, and remain there throughout postinfancy growth and development. This agrees with what is known about the growth of body fatness of children in developing countries who experience growth faltering (Eveleth & Tanner, 1990).

3. That the weight of organs other than the liver is not reduced in absolute terms when compared with the ideal male child shown in Table 5.4. As a consequence, the proportion of total body weight which comprises organ tissue is marginally greater than for the ideal child across postinfancy,

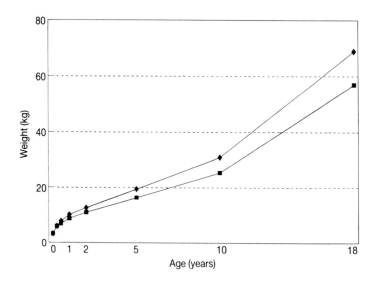

Fig. 5.4. Weights of the two children used in the model. One (diamonds) follows the fiftieth centile of height for age, National Center for Health Statistics (1977); the other (squares) shows growth faltering in late infancy and does not recover the deficit.

childhood and into adult life. Typically, only 6% of protein loss during starvation can be attributed to the liver (Uezu *et al.*, 1983); this value is incorporated into organ weight estimates at the ages when growth faltering is taking place.

4. Bone weight is estimated to be lower in direct proportion to lower body weight, and it has been estimated that the interspecies allometric relationship between bone weight and body weight has a power function close to one (Gunter, 1975). In the absence of any published values, it is assumed that the lower weight of viscera is also in direct proportion to lower body weight.

5. The proportion of body weight which is skeletal muscle is estimated by the difference between body weight and the sum of assumptions 1–4.

Comparing the ideal male following the fiftieth centile (Table 5.5) with the model male showing growth faltering in late infancy (Table 5.6), the latter has a greater proportion of body weight as organ tissue and skeletal muscle. In order to determine what influence this might have on BMR, theoretical BMRs are calculated from the weights and the known metabolic rates of the different tissues (Table 5.4). The relative contribution of brain, liver, heart and kidney to total organ weight across the growth period varies; the values used here are from Elia (1992).

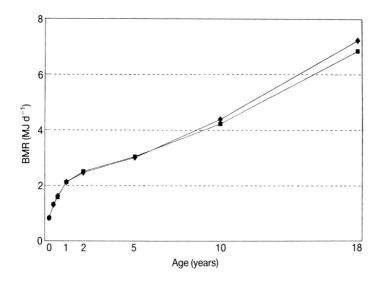

Fig. 5.5. Estimated basal metabolic rate across childhood of the fiftieth centile child (diamonds), and the one showing growth faltering in infancy (squares).

Figs. 5.5–5.7 show estimated BMR, BMR per kilogram, and BMR per kilogram FFM, respectively, for the fiftieth centile male and the male showing growth faltering. The BMR values for the fiftieth centile male are close to those reported by Schofield (1985) in his meta-analysis of BMR–body size relationships, for all ages apart from zero years. Thus the assumption made in the modelling procedure bear some resemblance to reality, at least for the normal child. The BMR of the child showing growth faltering is lower than that of the normal child (Fig. 5.5). However, when differences in body weight (Fig. 5.6) and fatness (Fig. 5.7) are taken into account, the reverse is true. This is entirely consistent with findings in India, where light, and light and short, adult males have lower BMR than normal controls, but greater BMR per kilogram and BMR per kilogram FFM (Kurpad, Kulkarni & Shetty, 1988). Similarly, a comparison of sleeping metabolic rates (SMR) (a proxy for BMR, since BMR under standardized conditions cannot be measured in the very young) of young Gambian and British children shows the former to have lower SMR (2.00 MJ d^{-1} against 2.37 MJ d^{-1}), but higher SMR per kilogram (226 kJ kg^{-1} d^{-1} against 205 kJ kg^{-1} d^{-1}) and SMR per kilogram lean body mass (LBM) (280 kJ kg^{-1} LBM d^{-1} against 259 kJ kg^{-1} LBM d^{-1}) (Vasquez-Velasquez, 1988).

In the studies of Kurpad *et al.* (1988) and Vasquez-Velasquez (1988),

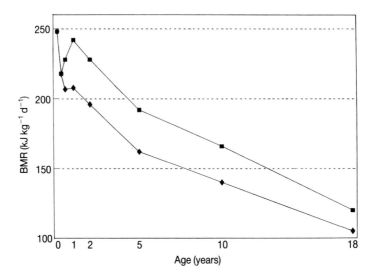

Fig. 5.6. Estimated basal metabolic rate per kilogram body weight across childhood of the fiftieth centile child (diamonds), and the one showing growth faltering in infancy (squares).

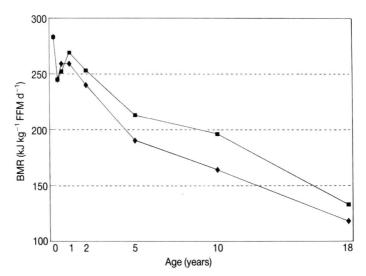

Fig. 5.7. Estimated basal metabolic rate per kilogram fat-free mass (FFM) across childhood of the fiftieth centile child (diamonds), and the one showing growth faltering in infancy (squares).

BMR and SMR are lower than predicted from body weight using the appropriate Schofield (1985) equations. For the Indian males, underpredictions of BMR are 4, 7 and 3% for the normal, light, and short and light groups respectively. These values are within the 95% confidence limits for prediction for the first and last groups, respectively. For the Gambian children, measured SMR is 5% lower than BMR predicted from body weight, while for the Cambridge, UK, children SMR is 9% lower. For both groups, the measured values are within the 95% confidence limits of prediction (Schofield, 1985), suggesting that there is no clear evidence of metabolic down-regulation in these two groups. However, populations with low BMRs are known to exist (Table 5.2); the modelling of body composition changes due to growth faltering in early childhood suggests that where BMRs lower than predicted exist, they are unlikely to be due to developmental plasticity.

Increased biochemical efficiency

Where BMR is lower than predicted, it can be postulated that metabolic efficiency at cellular level might be greater than expected. Waterlow (1986) proposed a number of possible ways in which such efficiencies might be obtained. The most important of these are reduced: (1) protein turnover; (2) ion-pumping, mostly sodium–potassium ATPase activity, but also calcium pumping; (3) futile cycling in energetically important metabolic pathways. Of these mechanisms, (1) and (2) are known to be responsive to thyroid hormone status (Wolman *et al.*, 1985; Kjeldsen *et al.*, 1984).

The production of T3 is low when the activity of the enzyme T4-5′-monodeiodinase is low (Waterlow, 1992). This iodinase is responsible for the generation of T3 from T4, a less active form of thyroid hormone, and is affected by the carbohydrate content of the diet, and by insulin. This is consistent with the observation that metabolic rate is increased with the ingestion of food (Garrow, 1978), and that circulating concentrations of T3 vary directly with the amount of dietary energy needed to maintain energy balance (Danforth, 1983), since both involve the ingestion of carbohydrate and the stimulation of an insulin response. Thus the lower than predicted BMR in some populations may be an indirect outcome of developmental plasticity, inasmuch as small body size due to growth faltering in childhood may require lower food intake to maintain energy balance. In turn, T4-5′-monodeiodinase activity would be reduced, T3 levels lowered, and BMR reduced as a consequence.

If the level of carbohydrate intake is of primary importance in this mechanism, then at any given body weight, individuals with higher levels of physical activity, maintaining energy balance with greater dietary intakes, would have higher T3 and BMR than individuals with lower levels of physical activity and eating less.

Low BMR is advantageous in that it conserves energy; it is likely to carry costs, too. In particular, a severe decrease in calcium ion-pumping could lead to accumulation of free calcium ions at intracellular sites, and associated cell damage (Jackson, 1990).

Plasticity and energetics

It seems unlikely that there is any direct plasticity in human BMR, although there may be plasticity in skeletal muscle fibre type. Most of the variation in human energy expenditure can be explained by variation in body size, but not body composition. The hormonal mechanisms operating in growth faltering may also serve to economize the energy expenditure of muscular activity, and to increase metabolic efficiency at the cellular level, giving lower BMR than expected for a given body size. Thus there are at least two ways in which plasticity in growth and energy expenditure can be related. Although merely suggestive, the evidence presented here points to the possibility that environmental influences on growth and development in childhood could have long-term energetic consequences that go beyond small body size.

References

Ashworth, A. (1968). An investigation of very low calorie intakes reported in Jamaica. *British Journal of Nutrition* 22, 341–55.

Belko, A. Z., Barbieri, T. F. & Wong, E. C. (1986). Effect of energy and protein intake and exercise intensity on the thermic effect of food. *American Journal of Clinical Nutrition* 43, 863–9.

Blaxter, Sir K. & Waterlow, J. C. (1985). *Nutritional Adaptation in Man*. London: John Libbey.

Danforth, E. (1983). The role of thyroid hormones and insulin in the regulation of energy metabolism. *American Journal of Clinical Nutrition* 38, 1006–17.

Danforth, E. (1985). Hormonal adaptation to over- and underfeeding. In: *Substrate and Energy Metabolism*, ed. J. S. Garrow & D. Halliday, pp. 155–66. London: John Libbey.

Danforth, E. & Burger, A. G. (1989). The impact of nutrition on thyroid hormone physiology and action. *Annual Reviews in Nutrition* 9, 201–27.

Elia, M. (1992). Organ and tissue contribution to metabolic rate. In: *Energy Metabolism: Tissue Determinants and Cellular Corollaries*, ed. J. M. Kinney & H. N. Tucker, pp. 61–79. New York: Raven Press.

Eveleth, P. B. & Tanner, J. M. (1990). *Worldwide Variation in Human Growth*. Cambridge University Press.

Frigerio, C., Schutz, Y., Whitehead, R. & Jecuier, E. (1992). Postprandial thermogenesis in lactating and non-lactating women from The Gambia. *European Journal of Clinical Nutrition* 46, 7–13.

Garrow, J. S. (1978). *Energy Balance and Obesity in Man.* Amsterdam: Elsevier Press.

Golden, M., Waterlow, J. C. & Picou, D. (1977). The relationship between dietary intake, weight change, nitrogen balance and protein turnover in man. *American Journal of Clinical Nutrition* **30**, 1345–8.

Goldspink, G. (1975). Biochemical energetics for fast and slow muscles. In: *Comparative Physiology – Functional Aspects of Structural Materials*, ed. L. Bolis, H. P. Maddrell & K. Schmidt-Nielsen, pp. 173–85. Amsterdam: North Holland Publishing Company.

Gunter, B. (1975). Dimensional analysis and theory of biological similarity. *Physiological Reviews* **55**, 659–99.

Hardeveld, C. van & Clausen, T. (1984). Effect of thyroid status on K^+-stimulated metabolism and ^{45}Ca exchange in rat skeletal muscle. *American Journal of Physiology* **247**, E421–30.

Henriksson, J. (1990) The possible role of skeletal muscle in the adaptation to periods of energy deficiency. *European Journal of Clinical Nutrition* **44** (supplement 1), 55–64.

Henry, C. J. K. & Rees, D. G. (1988). A preliminary analysis of basal metabolic rate and race. In: *Comparative Nutrition*, ed. K. Blaxter & I. Macdonald, pp. 149–59. London: John Libbey.

Henry, C. J. K. & Rees, D. G. (1991). New predictive equations for the estimation of basal metabolic rate in tropical peoples. *European Journal of Clinical Nutrition* **45**, 177–85.

Holliday, M. A. (1986). Body composition and energy needs during growth. In: *Human Growth: A Comprehensive Treatise*, ed. F. Falkner & J. M. Tanner, pp. 101–17. New York: Plenum Press.

Horton, E. S. (1983). Introduction: an overview of the assessment and regulation of energy balance in humans. *American Journal of Clinical Nutrition* **38**, 972–7.

Isley, W. L., Underwood, L. E. & Clemmons, D. R. (1983). Dietary components that regulate serum somatomedin-C concentrations in humans. *Journal of Clinical Investigation* **71**, 175–82.

Jackson, A. A. (1985). Nutritional adaptation in disease and recovery. In: *Nutritional Adaptation in Man*, ed. K. Blaxter & J. C. Waterlow, pp. 111–25. London: John Libbey.

Jackson, M. J. (1990). Intracellular calcium, cell injury and relationships to free radicals and fatty acid metabolism. *Proceedings of the Nutrition Society* **49**, 77–81.

Jolesz, F. & Sreter, F. A. (1981). Development, innervation, and activity-pattern induced changes in skeletal muscle. *Annual Reviews in Physiology* **43**, 531–52.

Jones, C. D. R., Jarjon, M. S., Whitehead, R. G. & Jecquier, E. (1987). Fatness and the energy cost of carrying loads in African women. *Lancet* ii, 1331–2.

Keys, A., Brozek, J., Henschel, A., Michelson, O. & Taylor, H. L. (1950). *The Biology of Human Starvation.* Minneapolis: University of Minnesota Press.

Kinabo, J. L. & Durnin, J. V. G. A. (1990). Thermic effect of food in man: effect of meal composition and energy content. *British Journal of Nutrition* **64**, 37–44.

Kjeldsen, K., Norgaard, A., Gotzsche, C. O., Thomassen, A. & Clausen, T. (1984). Effect of thyroid function on number of Na-K pumps in human skeletal muscle. *Lancet* ii, 8–10.

Kulkarni, R. N. & Shetty, P. S. (1992). Net mechanical efficiency during stepping in

108 S. J. Ulijaszek

chronically energy-deficient human subjects. *Annals of Human Biology* **19**, 421–5.

Kurpad, A. V., Kulkarni, R. N. & Shetty, P. S. (1988). Reduced thermoregulatory thermogenesis in undernutrition. *European Journal of Clinical Nutrition* **43**, 27–33.

Lerberghe, W. van (1989). Growth, infection and mortality: is growth monitoring an efficient screening instrument? In: *Auxology 88. Perspectives in the Science of Growth and Development*, ed. J. M. Tanner, pp. 101–10. London: Smith-Gordon.

Limas, C. J. (1978). Calcium transport ATPase of cardiac sarcoplasmic reticulum in experimental hyperthyroidism. *American Journal of Physiology* **235**, H745–51.

Maloiy, G. M. O., Heglund, N. C., Prager, L. M., Cavagna, G. A. & Taylor, C. R. (1986). Energetic costs of carrying loads: have African women discovered an economic way? *Nature* **319**, 668–9.

Millward, D. J. (1986). Hormonal responses to low intakes in relation to adaptation. In: *Proceedings of the XIII International Congress of Nutrition*, ed. T. G. Taylor & N. K. Jenkins, pp. 419–23. London: John Libbey.

Minghelli, G., Schutz, Y., Whitehead, R. & Jequier, E. (1991). Seasonal changes in 24-h and basal energy expenditures in rural Gambian men as measured in a respiration chamber. *American Journal of Clinical Nutrition* **53**, 14–20.

National Center for Health Statistics (1977). *NCHS growth curves for children, birth-18 years, United States*. (DHEW Publication No. (PHS) 78-1650.) Hyattsville, Maryland: United States Department of Health, Education and Welfare.

Nicol, C. J. M. & Bruce, D. S. (1981). Effect of hyperthyroidism on the contractile and histochemical properties of fast and slow twitch skeletal muscle in the rat. *Pflugers Archiv* **390**, 73–9.

Norgan, N. G. (1990). Thermogenesis above maintenance in humans. *Proceedings of the Nutrition Society* **49**, 217–26.

Nwoye, L. O. & Goldspink, G. (1981). Biochemical efficiency in intrinsic shortening speed in selected vertebrate fast and slow muscles. *Experientia* **37**, 856.

Pelletier, D. L. (1991). *Relationships between Child Anthropometry and Mortality in Developing Countries: Implications for Policy, Programs, and Future Research*. Cornell University: Cornell Food and Nutrition Policy Program (Monograph No. 12).

Prentice, A. M., Coward, W. A., Davies, H. L., Murgatroyd, P. R., Black, A. E., Goldberg, G. R., Ashford, J., Sawyer, M. & Whitehead, R. G. (1985). Unexpectedly low levels of energy expenditure in healthy women. *Lancet* i, 1419–22.

Quenouille, M. H., Boyne, A. W., Fisher, W. B. & Leitch, I. (1951). *Statistical studies of recorded energy expenditure of man. Part I. Basal metabolism related to sex, stature, age, climate and race*. Technical Communication No. 17. Bucksburn, Aberdeenshire: Commonwealth Bureau of Animal Nutrition.

Schantz, P., Henriksson, J. & Jansson, E. (1983). Adaptation of human skeletal muscle to endurance training of long duration. *Clinical Physiology* **3**, 141–51.

Schofield, W. N. (1985). Predicting basal metabolic rate: new standards and review of previous work. *Human Nutrition: Clinical Nutrition* (supplement 1), 5–41.

Shephard, R. J. (1991). *Body Composition in Biological Anthropology*. Cambridge

University Press.

Strickland, S. S. & Ulijaszek, S. J. (1990). Energetic cost of standard activities in Gurkha and British soldiers. *Annals of Human Biology* **17**, 133–44.

Suko, J. (1973). The calcium pump of cardiac sarcoplasmic reticulum. Functional alterations at different levels of thyroid state in rabbits. *Journal of Physiology* **228**, 563–82.

Uezu, N., Yamamoto, S., Rikimaru, T., Kishi, K. & Inoue, G. (1983). Contributions of individual body tissues to nitrogen excretion in adult rats fed protein deficient diets. *Journal of Nutrition* **113**, 105–14.

Ulijaszek, S. J. (1992). Human energetics methods in biological anthropology. *Yearbook of Physical Anthropology* **35**, 215–42.

Underwood, L., Smith, E. P., Clemmons, D. R., Maes, M., Maiter, D. & Ketelslegers, J.-M. (1989). The production and actions of insulin-like growth factors: their relationship to nutrition and growth. In: *Auxology 88. Perspectives in the Science of Growth and Development*, ed. J. M. Tanner, pp. 235–49. London: Smith-Gordon.

Vasquez-Velasquez, J. L. (1988). The energy metabolism of children. PhD thesis, University of Cambridge.

Viteri, F. E., Torun, B., Garcia, J. C. & Herrera, E. (1971). Determining energy balance of agricultural activities by respirometer and energy balance techniques. *American Journal of Clinical Nutrition* **24**, 1418–30.

Waterlow, J. C. (1986). Notes on the new estimates of energy requirements. *Proceedings of the Nutrition Society* **45**, 351–60.

Waterlow, J. C. (1988). The natural history of stunting. In: *Linear Growth Retardation*, ed. J. C. Waterlow, pp. 1–12. New York: Raven Press.

Waterlow, J. C. (1990). Mechanisms of adaptation to low energy intakes. In: *Diet and Disease in Traditional and Developing Societies*, ed. G. A. Harrison & J. C. Waterlow, pp. 5–23. Cambridge University Press.

Waterlow, J. C. (1992). *Protein Energy Malnutrition.* London: Edward Arnold.

Wendt, I. R. & Gibbs, C. L. (1973). Energy production of rat extensor digitorum longus muscle. *American Journal of Physiology* **224**, 1081–6.

Wiles, C. M., Young, A., Jones, D. A. & Edwards, R. H. T. (1979). Muscle relaxation rate, fibre-type composition and energy turnover in hyper- and hypo-thyroid patients. *Clinical Science* **57**, 375–84.

Wolman, S. J., Sheppard, H., Fern, M. & Waterlow, J. C. (1985). The effect of tri-iodothyronine (T3) on protein turnover and metabolic rate. *International Journal of Obesity* **9**, 459–63.

6 The study of migrants as a strategy for understanding human biological plasticity

GABRIEL W. LASKER

The strategy of using migration studies to estimate the degree of human plasticity was developed by Boas (1911) and utilized by Shapiro, who directed the collection of data for this purpose by William Lessa (Shapiro, 1931), Frederick Hulse (Shapiro, 1939) and Frederick Thieme (Thieme, 1957). The early history of the subject is covered in the introductions to various research reports and in the review by Kaplan (1954). The subsequent development of the strategy has also been chronicled (Baker, 1976; Little & Baker, 1988). It remains for me only to give an informal account of some of my own research experiences.

As an undergraduate I had little exposure to research. In the summer of 1930 or 1931 I completed my requirements at the Experimental College of the University of Wisconsin by writing a social survey of the city of Yonkers, New York. All I remember about it is that the streets had been planned on a map and when actually constructed some of them ended abruptly at the foot of hills too steep to be climbed except by stairs. These impediments to road traffic coincided with the boundaries of neighborhoods.

A few years later – probably 1933 – a professor of international economics at the University of Michigan set me to study the rice trade of Indo-China. I had great trouble meeting the language requirement for graduation, but I enjoyed the same privilege as graduate students to work in the library stacks on French-language colonial reports.

After graduation from the University of Michigan in 1934 I taught myself human genetics and did the genetic analysis of an inborn error of metabolism that my mother, Margaret Lasker, and Morris Enklewitz had been studying from the biochemical and clinical points of view (Lasker, Enklewitz & Lasker, 1936). However, it was in the Great Depression and I could not get a proper job in the United States, so I went to China in 1935 and studied Mandarin in Peking. (I use the English spelling for the same reason that I use the spellings 'Vienna' and 'Rome'.) Otto Klineberg, the distinguished social psychologist, and his wife were among the other residents at the College of Chinese Studies where I lived for the first year of

110

my stay in China. Klineberg had been a student of Franz Boas, among others; in his doctoral thesis project, Klineberg applied a version of Boas' migration strategy to the study of intelligence of 'Negroes' in the South and migrants from there to Harlem. I read and reread his book that summarized these studies and wrote and published a review of it (Lasker, 1937).

Paul Stevenson was professor of anatomy at Peking Union Medical College. He had been a student of T. Wingate Todd, an anatomist with anthropological interests. Stevenson guided me in a library study of Chinese physiological variables such as basal metabolism and blood pressure and the changes in these variables in Chinese migrants to the United States. The part of my study which dealt with growth of Chinese children eventually appeared in print in China four years later in the middle of the war with Japan (Lasker, 1941). In the meanwhile, Stevenson, Klineberg and others strongly recommended me and, despite my abysmal undergraduate record, Professor Hooton accepted me into the graduate program in physical anthropology at Harvard.

Hooton's chief interest was in racial types, but he granted his students great leeway and support and allowed me to include the question of changes in milieu through migration in my PhD study of Chinese in America (Lasker, 1946). I went straight to Boas and Shapiro at Columbia University for advice. Both spoke of difficulties with such a study and Boas did a quick paper and pencil power analysis to show that my results would be unlikely to be significant. His prediction later proved largely correct and the study, although important in my education, was too small for conclusive results. Furthermore, the conclusions were weak because I was unable adequately to allow for the interaction of anthropometric measurements and age. On average, the immigrant Chinese born in China were much older than the American-born of Chinese descent. Shapiro (1931) had previously called attention to the fact that most Chinese immigrants to Hawaii had come before the passage of the Chinese exclusion act and were therefore much older than most Hawaii-born Chinese and that it is difficult to allow for this by correlation because the curves of distribution of ages are not normal. I was unable to surmount this difficulty with the methods available at the time of my study. The conclusions, although not rigorous, met the standards of the time and are probably reasonably acceptable. Shapiro, with his more rigorous standards, never satisfied himself enough to publish more than the preliminary report of the data collected for him by Lessa.

On my frequent visits to Columbia University, I went not only to the anthropology department (where Weltfish, Benedict, Boas and Strong treated me as if I were one of their own students) but also to the psychology department housed in the same building. Two of Klineberg's students of psychology (Robert Chin and Herbert Hyman) often accompanied me to Chinatown where I measured some of the subjects of my study. The informal discussions with these thoughtful social psychologists helped prepare me to expect biological variables to be plastic in a way that most physical anthropologists of the time rejected in their studies of anthropometric traits. The intellectual climate at Columbia University influenced my choice of the hypothesis for my thesis that the interactions between genetics and environment manifest different degrees of plasticity for different variables, but are always an element in the ecological equation.

Our study in Mexico, beginning in 1948, followed the work there of Marcus Goldstein, one of Boas' students. Goldstein (1943) had studied sedente Mexican families in Mexico and migrant families and their American-born offspring in Texas. Goldstein's original case-by-case data sets were published and remain a resource for the multivariate analyses that the complex research plan deserved but that were impracticable when the work was done.

Shapiro (1939) in his study of Japanese migrants had ascribed the differences between sedentes in Japan and migrants to Hawaii to selective migration. That interpretation was probably incorrect; Hulse, who had done the fieldwork for the study, once told me that some of the migrants had come to Hawaii at a young age when their growth could still have been influenced. We were able to test the question by a study of returned emigrants and sedentes in Mexico (Lasker, 1952; Lasker & Evans, 1961). These studies showed that first-generation migrants themselves, if they migrated while young, manifested a plastic response to the changed environment; that is, those who migrated when fully adult were not significantly different from the sedentes, but those who had migrated when younger were in some ways more similar to the American-born. Furthermore, one small prospective study showed virtually no selective migration for the measurements usually included in such studies (Lasker, 1954).

One other factor may influence the interpretation of studies of migrants. In general migrants are more likely than sedentes to mate exogamously, hence with individuals genetically unlike themselves (i.e. heterogamously). Some of the anthropometric difference between first-generation migrants and the second generation (e.g. American-born children of immigrants) may thus be due to a so-called heterosis (genetic 'hybrid vigor') rather than to direct influence of the environment on growth (Hulse, 1957). However, even in Hulse's study this could account for only a fraction of the difference recorded. The effect of such a factor is small or non-existent in some studies and, overall, is probably small or insignificant (Lasker, Kaplan & Sedensky. 1990).

Previous chapters refer to my review article concerning the levels of adaptability (Lasker, 1969). The background for that review was that the Human Adaptability Project of the International Biological Program was being organized and, as a non-participant and non-applicant for funds, I was asked to review the status of the subject. I doubt that it was an original observation, but one point I emphasized was the variation in time it takes to make adaptations: (1) minutes or days for many physiological variables; (2) one generation for traits fixed by growth; and (3) many generations for genetic characteristics through Darwinian selection. The more a plastic response is developed in the species, the less the strength of the vectors for slower forms of adaptation. That is, the evolution of rapid plastic responses tends to inhibit slower modification that requires one or many generations.

The question has been raised as to whether there is such a thing as plasticity in the narrow sense of changes during the growth period becoming permanent and adaptive throughout adulthood. Certainly a question remains concerning any advantage to adults of some of the kinds of environmentally mediated modifications cited above. There remain examples for which there can be little dispute, however. For instance, the immune system can respond permanently and advantageously. A person who has once had measles does not acquire the disease again if subsequently exposed.

The immune response is sometimes, by analogy, referred to as 'learned'; learning itself is an example of plasticity that is particularly well developed in humans. Once an individual learns to make or use a tool, select edible from inedible plants, or communicate by spoken word, the memory is a more or less permanent change in the brain which may last for his or her lifetime, but it is not transmitted genetically. Thus learned behavior is the chief form of biologically adapting individuals that occurs during development, is essentially irreversible, but is not imbedded in the population by natural selection of genotypes. Whatever the case for plasticity in other anatomical structures, some changes in the brain give images and ideas that are 'hard to forget' and thus meet the narrow definition of the term 'biological plasticity'.

Today there is resurgent nationalism based on racism (the false notion that all of what one can do is predominantly imbedded in genes derived from a circumscribed biological race). Therefore a full understanding of the role and extent of biological plasticity by which the messages of DNA are modified in the anatomy and behavior of people deserves wide dissemination. I congratulate the authors of the other chapters in this volume for expanding knowledge about plasticity in the broadest sense. Their eloquence derives from the persuasiveness of the results of their studies and will, to some extent, be measured by understandings that you, the readers, carry away with you and help to promulgate.

References

Baker, P. T. (1976). Research strategies in population biology and environmental stress. In: *The Measures of Man: Methodologies in Biological Anthropology*, ed. E. Giles & J. S. Friedlaender, pp. 230–59. Cambridge, Massachusetts: Peabody Museum Press.

Boas. F. (1911). *Changes in Bodily Form of Descendants of Immigrants.* Washington: The Immigration Commission, Government Printing Office.

Goldstein, M. S. (1943). *Demographic and Bodily Changes in Descendants of Mexican Immigrants.* Austin, Texas: Institute of Latin American Studies, University of Texas.

Hulse, F. S. (1957). Exogamie et hétérosis. *Archives Suisse d'Anthropologie Général* **22**, 103–25. (Translated into English with a comment by G. W. Lasker as: Exogamy and heterosis. *Yearbook of Physical Anthropology* **9**, 240–57.

Kaplan, B. A. (1954). Environment and human plasticity. *American Anthropologist* **56**, 780–800.

Lasker, G. W. (1937). [Book review.] Otto Klineberg. *Race Differences.* Nankai Quarterly, Tientsin, China. [No page numbers given on the available reprint.]

Lasker, G. W. (1941). The process of physical growth of the Chinese. *Anthropological Journal of the Institute of History and Philology, Academia Sinica* **2**, 58–90.

Lasker, G. W. (1946). Migration and physical differentiation. A comparison of immigrant and American-born Chinese. *American Journal of Physical Anthropology* **4**, 273–300.

Lasker, G. W. (1952). Environmental growth factors and selective migration. *Human Biology* **24**, 262–89.

Lasker, G. W. (1954). The question of physical selection of Mexican migrants to the U.S.A. *Human Biology* **26**, 52–8.

Lasker, G. W. (1969). Human biological adaptability. *Science* **166**, 1480–6.

Lasker, G. W. & Evans, F. G. (1961). Age, environment and migration: Further anthropometric findings on migrant and non-migrant Mexicans. *American Journal of Physical Anthropology* **19**, 203–11.

Lasker, G. W., Kaplan, B. A. & Sedensky, J. A. (1990). Are there anthropometric differences between the offspring of endogamous and exogamous matings? *Human Biology* **62**, 247–9.

Lasker, M. Enklewitz, M. & Lasker, G. W. (1936). The inheritance of 1-xyloketosuria (essential pentosuria). *Human Biology* **8**, 243–55.

Little, M. A. & Baker, P. T. (1988). Migration and adaptation. In: *Biological Aspects of Human Migration*, ed. C. G. N. Mascie-Taylor & G. W. Lasker, pp. 167–215. Cambridge University Press.

Shapiro, H. L. (1931). The Chinese population in Hawaii. (Preliminary Paper Prepared for the Fourth General Session of the Institute of Pacific Relations.) New York: American Council, Institute of Pacific Relations.

Shapiro, H. L. (1939). *Migration and Environment.* New York: Oxford University Press.

Thieme, F. P. (1957). A comparison of Puerto Rican migrants and sedentes. *Papers of the Michigan Academy of Science, Arts, and Letters* **42**, 249–56.

7 Human migration: effects on people, effects on populations

D. COLEMAN

Summary

The term 'plasticity' is not used in demography but the concepts are, although in an analogous fashion and not in any direct physiological sense. Physiological effects upon individuals tend to be rather unimportant as demographic variables. This paper discusses two demographic phenomena relating to immigration which show obvious parallels with Lasker's concepts of 'plasticity'. Populations can respond, in a plastic or other fashion, to changes in their equilibrium conditions in at least two sorts of ways. One is for individuals to change their behaviour in response to new circumstances. These changes, which in the aggregate may be described as transition or acculturation, may have 'plastic' qualities of irreversibility, although the process is psychological and psychosocial, not physiological. The other relates to population dynamic responses to changes in equilibrium.

The first example describes the responses of individual immigrants, measured by their population trends in fertility and interethnic marriage in the 'host' countries to which they have moved. In general, immigrants from high-fertility demographic regimes and from traditional closed societies do moderate their birth rates, although to an extent which varies considerably according to the society of origin. Changes between generations are typically faster than those within generations.

At a 'macrodemographic' level, populations respond slowly to changes in their underlying vital rates and to immigration. These changes are not truly plastic, being reversible, but they are slowed by the inertia built into population age structures and always tend to return to an equilibrium defined by the underlying vital rates. These population responses to changes in immigration and fertility are important because they define the principles behind the contemporary debate on the ageing of Western society following the reduction of birth rates, and the possible beneficial effects of higher levels of immigration. Scenarios presented here illustrate the principle, already well known in demographic theory, that fertility has, other things being equal, more potent effects upon population age structure than does immigration, and that although immigration may affect population growth it cannot, unlike changes in fertility, arrest population ageing.

Introduction

Demographic, as opposed to genetic, consequences of migration may seem to be an unpromising contribution to a volume on human variability and plasticity. It must be confessed straight away that demographers are not in the habit of using the term or the concept of 'plasticity'. Neither do they pay much attention to the statistical variability of the phenomena that they describe. Much demography, dealing with whole populations rather than samples, concentrates on summary indices describing population behaviour and their differences between subgroups. This, too, is regrettable, as noted elsewhere (Coleman, 1995). In modern societies with low fertility, group differences in society such as social class or educational attainment typically account only for about 10% of the very considerable level of variation between individuals in terms of completed family size and the timing of births (Ni Bhrochain, 1993). The major exception is ethnic or immigrant origin, a topic to which I will return.

Although the primary subject matter of demography – reproduction and death – is biological, the variability of demographic responses are not primarily physiological in nature or causation. Demography measures the pattern and variation of the risks of membership of a particular society. In any given society these risks may have a characteristic statistical profile known as a demographic regime (in respect of the timing and number of births, the form of marriage, family and household, causes and level of mortality, etc.). These regimes vary considerably between different societies and types of society. The shape of this profile is a response to the economic environment and social structures which that society has created. In societies which are to survive, the demographic regime must be capable of reproducing the members of that society into their next generation. Its form is often thought therefore to be functionally adapted to those socioeconomic circumstances.

Physiological effects on individuals tend to be rather unimportant as demographic variables. Human fertility is considered to be determined by a number of necessary 'proximate' determinants which affect conception, fertilization, gestation and delivery (Bongaarts & Potter, 1983; Leridon & Menken, 1982). These determinants, including exposure to sexual intercourse, breast-feeding and other factors affecting fecundity, contraception and abortion, moderate the potential of reproduction (fecundity) into actual measured reproductive performance (fertility). Infectious disease aside, physiological impacts of, for example, nutritional status, appear to have (contrary to earlier claims (Frisch, 1978)) relatively modest effects on fertility compared with differences in individual behaviour and social arrangements relating to breastfeeding, contraception, age at marriage, etc. (Menken, Trussell & Watkins, 1981). Starvation, indeed, appears to have had relatively little effect even upon mortality in most populations in the past (Livi-Bacci, 1991). Nutritional status in general, although controversial, normally emerges as a minor component in the observed variation of human reproductive performance. It takes a lot of malnutrition or fat loss substantially to slow down, or to shut down, the human reproductive system (Lunn, 1988).

Plasticity and adaptibility at the population level
Can we usefully talk, therefore, about 'plasticity', 'adaptability' and other phenomena at the population level? Any such suggestions must be at the level of analogy. Populations can respond to changes in their equilibrium conditions in at least two sorts of ways. One is for individuals to change their behaviour in response to new circumstances. Marriage and births may be postponed or brought forward, individuals may migrate. These can be short-term adjustments to preserve the status quo, which might not, for example, change the final completed family size. More interesting adjustments come from a process of social change or transition where ideas and targets are altered by some new experience, as when people move from one socioeconomic system with a characteristic demographic regime to another where some of their demographic characteristics (possibly high fertility or complex household structure) does not fit in with new ways of earning a living or with available housing. These changes in the course of individual lifespans, which in the aggregate may be regarded as a transition or acculturation, may have certain 'plastic' qualities of irreversibility, even though the process involved is psychological and psychosocial, not physiological.

In demography the best known of these processes is generally known as the demographic transition, whereby populations move from demographic regimes characterized by high death and high birth rates, with minimal levels of population growth, through an intervening period of falling mortality and static or even rising fertility, generating transient but rapid population growth, to a new equilibrium of low birth and death rates with very low or even negative population growth rates (Chesnais, 1993).

Populations can also adjust at a more technical level. Population dynamics has rules, which have something in common with those of population genetics, which govern the equilibrium states of population age structures and the way in which they can return to equilibrium following changes in death rates, birth rates and migration, while acquiring a new population age-structure generated by the balance between the vital rates (McFarland, 1969). These rules, described by stable population theory, generate equilibrium positions which relate age and sex distributions in populations to that population's underlying birth and death rates. They are analogous to the Hardy–Weinberg equilibrium in population genetics or to Newton's laws of motion. In brief, the equilibrium (self-renewing) age-structure depends entirely upon the (constant) values of the underlying rates of fertility, mortality and migration (see Keyfitz, 1977). However, because of the inertia inherent in population age-structures, some decades must elapse before a new self-renewing age structure is established consequent upon new vital rates.

Purpose of this chapter

This chapter will look at adaptable responses, by large populations and by their members, to the effects of migration in these two ways: in terms of population characteristics and changes in individual behaviour, and in terms of population structure. It is particularly appropriate to choose migration in a volume celebrating the contributions to human biology of Gabrial Lasker, because his works, especially in population biology, have included seminal work on the effects of migration on population structure and especially on their relationship to the distribution of surnames (e.g. Lasker, 1985; Lasker, Mascie-Taylor & Coleman, 1986). He has made substantial contributions to the use of surnames in the estimation of population structure, and the effects on that structure of migration, in relation to both marriage and other phases of mobility.

The populations studied will not be the usual subjects of biological anthropological investigations. Biological anthropology has traditionally concentrated its interest upon small-scale populations or subpopulations, individual assessments of kinship and bioassay on relatively small samples of people. This is entirely appropriate given the likely effective size of human populations over most of human history, and is the appropriate population size for studying the effects of drift and founder effects, and their counteraction by migration, which can be the predominant modes of genetic change in such societies.

Demographic interest differs from the anthropological in that the populations studied are often between five and seven orders of magnitude bigger than those that interest anthropologists. This typically eliminates random sampling effects and stochastic processes as important factors in whole-population variation. Demography also typically operates over a much shorter time-frame than does anthropology. This reverses the relative importance of effects of social and biological characteristics in determining events. In demography, on the whole, the ultimate factors of change affecting the biological variables of birth and death rates are environmental, economic, social, ideational and political. They are not usually themselves biological in nature. Varied demographic regimes in different populations have different reproduction rates because of their social, economic or cultural settings, not because they have different biological or genetic constitutions.

Genetic consequences may flow from these arrangements, however. Populations in different parts of the world have started, or completed, the demographic transition at different points in time. In addition, some began with different levels of fertility and experienced different growth rates during the population growth phase of the transition. Consequently there have been rapid changes in the twentieth century in the relative numbers of human populations from different geographical regions and hence of different genetic and racial makeups. This shift in balance, towards people of African and Asian origin and away from people of European origin, although only half completed, is substantial and is likely to be permanent. Similar shifts in population origins occur at the national level when immigrant populations have different demographic regimes from natives. Change in human gene distribution today is primarily a passive consequence of such very large-scale demographic differences in growth rates, most of which have had cultural or socioeconomic, rather than biological, causes.

New minorities in Western Europe and their demographic adaptability

I will take as the first example the relatively recent rise of immigrant populations from Third World countries. These have brought unfamiliar demographic regimes, regional cultures and languages to Europe and to North America in the last few decades. The high fertility and youthful age-structure of these populations are rapidly changing population composition in many areas. I will concentrate on the European experience, as this may be less familiar to most readers.

Unlike the US, no Western European country regards itself as a 'country of immigration', especially Germany. Almost all appear to be becoming, in fact, 'countries of immigration', particularly Germany. Much of the increase in migration to Europe since the mid-1980s is by asylum claimants or 'ethnic' return migrants, particularly to Germany, or by illegal immigrants from the East and from the South. At the same time migration streams (labour migration, family reunion) which, in the early 1980s, had been thought to be under control or declining, have revived, and family formation migration of fiancés and spouses from the third world has increased. Almost all countries in Europe except Ireland and Portugal have experienced these changes, even the Mediterranean countries which used to send workers to Germany. Eastern European countries have come under immigration pressure for the first time, from the Commonwealth of Independent States (CIS) and from immigrants from the South using Eastern Europe as an easy back door into the West. Germany herself has taken the lion's share of asylum claimants: her generous asylum laws, only reformed in 1993, and her powerful economy, have proved irresistible to those seeking easy entry to Europe and a better life.

Immigration data: a health warning

It is necessary to begin with a health warning about immigration data. The volume of data on immigration may seem impressive but the quality is often poor. Comparable facts are in short supply. The huge scale of international travel makes it difficult to measure that small fraction of all movement which should be considered as 'international migration'. People move between countries for variable lengths of stay and for many different purposes. The United Nations definition of an international migrant is a person who enters another country for at least 12 months having been absent from it for at least 12 months. In all Europe only the UK provides data that match that criterion, and these only from a sample survey of passengers. Different national systems, recording movement defined by their own unique immigration laws, often produce incompatible statistics. For example, 208 000 of the immigrants to Western Europe in 1989 were labour migrants entering with work permits. But these figures come from just the six European countries which publish figures; only four of them are from the European Community (EC). Adding together data from different countries can be like adding apples and pears. Data on the numbers of immigrants entering a particular EC country from another EC country seldom match the numbers recorded as leaving the latter country to settle in the former (Poulain, 1993). All migration streams produce a counterstream back to the destination, but in democracies that do not control exit, emigration data tend to be weaker than immigration data. Controls are substantially evaded by illegal immigrants, and national control systems are poorly funded with particular neglect of modern information technology systems.

Flows of migrants into Europe

Each year since the late 1980s at least 2 million immigrants, including asylum seekers, have entered Western European countries, about 5 immigrants per 1000 population per year. These figures include labour migrants, the dependents or spouses of previous immigrants entering for family reunification, and the exceptional inflow of German 'ethnic migrants' or *aussiedler* to Germany from Eastern Europe (377 000 in 1990) (Table 7.1). People also leave. Emigration flows are less well counted, but can be a third or more of the inflow in relation to Third World countries, and in relation to industrial countries the inflow and outflow may be equal or negative. In 1990 and in 1991, about 1.4 million foreign immigrants are known to have entered Western European countries legally, excluding asylum seekers. Net legal immigration in 1990 and 1991 was about half the gross figure. The total of entrants is increased to about 2 million or more each year by the large numbers of asylum seekers (426 000 in 1990; 544 000 in 1991; 680 000 in 1992), most of whom remain in Europe, often illegally, even though the asylum claims are usually rejected. All these categories of migrants, legal, asylum and illegal, have been increasing since the mid-1980s.

Stock of foreign population in Europe

As a result of this growing influx, the Western European countries have a bigger and more diverse immigrant or foreign-origin population than ever (Table 7.2). In 1991 there were about 15.2 million legally resident foreign citizens in the EC countries, in addition to several million former foreign citizens who have become naturalized. The rate of naturalization in France is such that the number of foreigners counted in the census of 1990 was slightly less than in 1982, despite eight years of immigration. Of the foreigners in the EC in 1991, 5.0 million were citizens of other EC member states and another 1.9 million from other European countries; 8.3 million were citizens of non-European countries, mostly from the Third World, over 2 million each from Turkey, from Africa including the Maghreb (Tunisia, Morocco, Algeria), from Asia and the Americas. Thirty-six per cent of the total number of foreign residents and more than half of the non-EC nationals come from the Maghreb, Turkey and Yugoslavia. In 1991 foreign population ranged from 17% of the total population in Switzerland, to 9% in Belgium, 8% in Germany and 7% in France. In Spain and Italy only about 1% of (legal) residents are foreign but the numerous illegal immigrants in those countries will at least double that proportion. In absolute numbers, Germany has the largest number of foreigners of any European country (6 million in 1992). It will be noted that the flows of immigrants into these 'non-immigration countries' are in many cases considerably higher than those into the US in relation to population size, and that the proportions of foreign population are about the same in many cases (although the totals in Table 7.1 (landscape)

Table 7.1. *Immigration into nine EC countries, 1991*
Note: these totals are not wholly compatible. The European totals include some asylum seekers.

	Population size (thousands)	All immigrants	EC citizens	Non-EC citizens	All immigrants[a]	EC citizens[a]	Non-EC citizens[a]
Belgium	9 987	87 480	38 186	49 294	8.8	3.8	4.9
Denmark	5 147	46 587	26 110	20 477	9.1	5.1	4.0
Germany	79 753	1 182 027	380 790	801 237	14.8	4.8	10.0
Greece	10 057	34 348	13 972	20 376	3.4	1.4	2.0
Spain	38 994	24 320	17 047	7 273	0.6	0.4	0.2
France	56 652	102 109	8 320	93 789	1.8	0.1	1.7
Italy	57 746	126 835	82 050	44 785	2.2	1.4	0.8
Netherlands	15 010	120 249	55 907	64 342	8.0	3.7	4.3
UK	57 000	267 000	148 000	119 000	4.7	2.6	2.1
Total	330 347	1 990 955	770 382	1 220 573	6.0	2.3	3.7
Immigration flows into 'Countries of Immigration' 1990 (for permanent settlement)							
Australia	17 085	122 000			7.1		
Canada	26 452	214 000			8.1		
USA	251 142	656 000			2.6		

[a]These figures are immigrants per thousand of the population. They represent gross, not net, flows.
Source: Eurostat 1993 (unpublished); national demographic yearbooks.

Table 7.2. *Foreign population* 1990, *selected countries*

Population figures are in thousands.

	All	From EC	National population	All as percentage
Belgium	905	552	9987	9.1
France	3597	1312	56652	6.4
Germany	5518	1439	79753	6.9
Netherlands	692	169	15010	4.6
Norway	143	41	4233	3.4
Sweden	456	65	8527	5.3
Switzerland	1127	760	6674	16.9
UK	1791	781	57323	3.1
E&W[a] ('NC ethnic')	3003	—	50900	5.9
Foreign birthplace around 1990, selected countries				
USA (1990)	19800	—	249714	7.9
Canada (1986)	3908	—	25022	15.6
Australia (1991)	3941	—	17336	22.7

[a] E & W Preliminary data, 1991 census.
Sources: EC (1992), Council of Europe (1992), national demographic yearbooks or census, Population Reference Bureau Inc.

Europe are underestimated by the high rates of naturalization in some countries).

These foreign populations continue to grow, at between 1% and 6% per year, through a combination of continued immigration and high natural increase. Naturalization, on the other hand, tends to reduce the numbers of foreigners although not, of course, the population of foreign or immigrant origin. Fertility of foreign populations in Europe is generally declining. Among some populations, Indians in the UK and people of Caribbean origin in the Netherlands and the UK, fertility is about the same as the national average, although thanks to a youthful age-structure the number of births still increases. In other cases, especially among Turks and some other Muslim populations, fertility remains equivalent to a family size of between 3 and 5 children (Coleman, 1994). The proportion of all births to foreign mothers ranges from 7% in the Netherlands, 10% in Sweden, 11% in France and Germany and 12% in the UK (immigrant mothers). Whether these children will also be counted as foreigners depends on the citizenship laws of each country.

Adaptation of immigrant demographic regimes: an accelerated transition?

There is considerable interest in the adaptability of new immigrant demographic regimes to their new socioeconomic environments. Do immigrant fertility regimes adapt by declining to host population levels in an accelerated demographic transition? Or will pressures to preseve old traditions of in-marriage, high fertility and restrictive roles for women prevail, as a kind of 'defensive structuring' to protect the existence of the traditional community and prevent its absorption by the wider society (Siegel, 1970)? A number of theoretical possibilities exist, including the possibility that immigrant minorities may experience an accelerated demographic transition as an aid to maximizing material success in a socially unwelcoming but economically helpful environment, It is, however, by no means certain that the fertility of minorities will converge on that of the host population. Persistent economic differences and/or cultural characteristics may prevent that happening. For example, in the US the minorities of Chinese and Japanese origin have preserved fertility levels consistently lower than the US average over many decades (Bean & Frisbie, 1978), even though the fertility levels in the home country were higher than those of the US over most of the period. Over a much longer time period, the birth rate, (measured as the total fertility rate (TFR)) of the US black population has failed to converge with that of whites, even though both clearly follow the same trends (Fig. 7.1). The process may be likened to an enormous if unintended experiment, whereby the demographic regimes of diverse human populations (subjects) have been subjected to the effects of a variety of alien cultural, economic and political influences in the European host societies (treatments) where they have come to live.

The first wave of labour migrants to the rapidly growing economies of Germany and other NW European countries in the 1960s came from the (then) poorer southern European countries of Portugal, Spain, Italy, Greece and, later, the former Yugoslavia. The cultural distance was not great and the fertility levels were generally not very elevated. None the less the trend has generally been for the fertility levels of the immigrant population to depart from the higher levels of the country of origin and converge upon those of the host society. In a number of cases, interesting changes then take place. In many cases, among Spanish immigrants in Germany (Fig. 7.2) and Yugoslavs in Sweden (Fig. 7.3) the fertility of immigrants declines to about that of the host country and then follows the trend of the host country. This accommodation to the host country norms has persisted even when the fertility trends in the home country have fallen below those of the sending countries. Italy and Spain now have the lowest

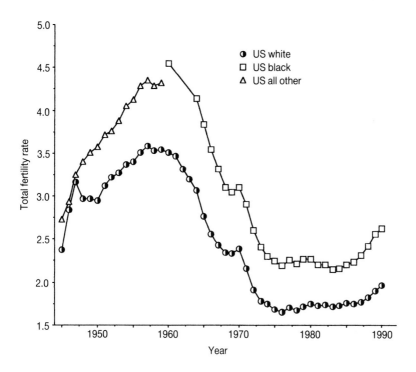

Fig. 7.1. Total fertility rate by race, United States 1945–90. (Sources: NCHS (1993) Monthly Vital Statistics Report, vol. 41, no. 9 (Supplement): *Advanced Report of Final Natality Statistics 1990*, table 4; NCHS (1977) Vital and Health Statistics Series 21, no. 28, *Trends in Fertility in the United States*, table 13; NCHS (1990) *Vital Statistics of the United States 1988*, vol. 1 (*Natality*) (PHS), pp. 90–1100, tables 1–6.)

fertility (under 1.3) of any major country in the world, and Greece and Portugal are not far behind.

Even immigrant groups that retained higher fertility relatively late, like the Portuguese in France, now have birth rates (1.9 in 1990) little more than the national average. The pattern is not completely clear-cut. Greek fertility in Germany has fallen below the German national level. Furthermore, immigrant populations from the same origin may respond somewhat differently in different host populations. The TFR of Italians in Germany is somewhat higher than the German national average (1.7 in 1987) and considerably higher than the figure for Italian immigrants in France (1.4 in 1990), which is well below the French average. These data, it must be noted, are mostly based upon data of births to women of foreign nationality. The base population and its composition therefore changes in response to the

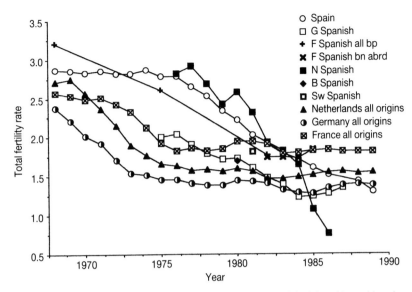

Fig. 7.2. Total fertility rate trends of women of Spanish citizenship resident in Germany, France and The Netherlands 1974–89, compared with home countries. (Source: Coleman (1994).)

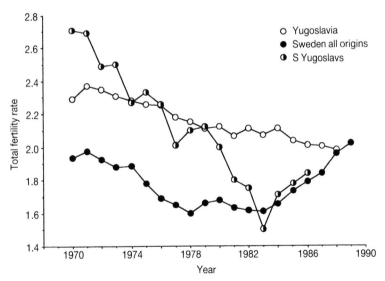

Fig. 7.3. Total fertility rate trends of women of Yugoslav citizenship resident in Sweden, 1970–86, compared with home countries. (Source: Coleman (1994).)

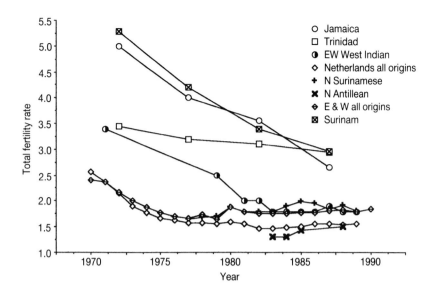

Fig. 7.4. Total fertility rate trends of women born in the West Indies and resident in England and Wales, and of Surinamese and of Antillean origin resident in the Netherlands, 1970–87, compared with home countries. (Source: Coleman (1994).)

pace of immigration and the rules of naturalization and also to national changes in nationality law, particularly with respect to the changes in the law concerning the automatic acquisition of nationality by spouses of nationals.

With non-European immigrant groups, the trends have not been so simple. Most Third World immigrants came from populations that had, and in many cases still have, considerably higher levels of fertility than those customary in the West. None came with the West European system of marriage and household. Instead, all their regimes were based either upon various forms of extended family households, where marriage was universal, or in Caribbean populations, a very different form where marriage was late and births outside marriage frequent. Both in the UK and in the Netherlands, Caribbean populations have been relatively unsuccessful economically. But Surinamese and Antilleans in the Netherlands and West Indians in the UK now have among the lowest fertility rates of any non-European immigrant group, scarcely distinguishable from those of the host populations and in the case of the Antilleans, actually below it (Fig. 7.4). These data are difficult to judge in the UK case, as the figures refer to immigrant women only and most West Indians of peak reproductive age in

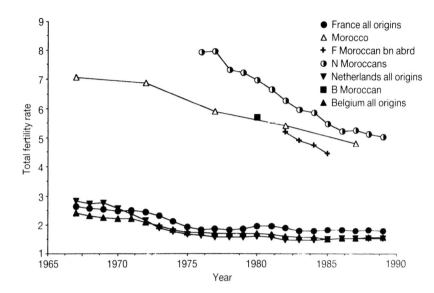

Fig. 7.5. Total fertility rate trends of women of Moroccan citizenship resident in various European countries 1968–89, compared with home countries. (Source: Coleman (1994).)

the UK were themselves born in the UK. However, the correspondence with the Dutch data is supportive. It is unfortunate, for these purposes, that the French population of West Indian origin (perhaps 50 000 people from Martinique and Guadeloupe) are all French citizens and are therefore not distinguished in French statistics.

By contrast, the immigrant population from the Maghreb have shown much slower fertility declines. Some, indeed, had birth rates in Europe higher than in the country of origin, although it is important to remember that the TFR statistic can be inflated when family formation patterns are disrupted by the processes of migration and of delayed family reunification. Moroccans in the Netherlands and in France have shown a rate of fertility decline little faster than in the demographic transition in the Maghreb itself (Fig. 7.5). Tunisians in France show similar trends to Moroccans, but the fertility of Algerians had fallen lower, to 4.0 in 1985 and 3.2 in 1990 (Isnard, 1992). Turks in Germany, Sweden, France and the Netherlands share common trends. Most began with a birth rate higher than in Turkey itself (where fertility is more moderate than in the Maghreb). Fertility in Europe fell until the mid-1950s and then appeared to be rising again. It is not yet known why, but it may be that a higher proportion of foreign mothers are relatively recent brides from Turkey, who have not been brought up

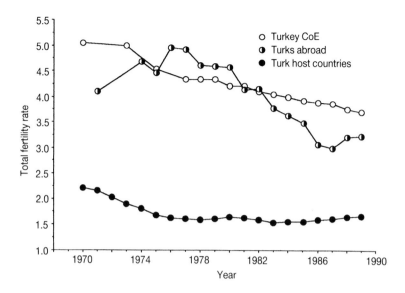

Fig. 7.6. Average total fertility rate trends of women of Turkish citizenship in various European host countries, 1970–89, compared with home countries. Host countries include France, Germany, The Netherlands and Sweden. (Source: Coleman (1994).)

exposed to Western norms of family size or contraception (Fig. 7.6).

Asians in Britain provide an important test, as they include one of the few substantial Asian immigrant populations which is not predominantly Muslim. The East African Asians in Britain, mostly Hindu, arrived with an already favourable socioeconomic background, having provided the middle class for East Africa's economies before their expulsion in the process of 'Africanization'. Their fairly rapid fertility transition is therefore not very surprising. The Indian population from India is even more interesting as, like most immigrants from South Asia, many came from relatively humble backgrounds. This Indian population is of course varied ethnically and linguistically (Robinson, 1986) but most are not Muslims. They have achieved a relatively rapid social mobility (Robinson, 1988), so that together with the East African Asians they have a higher socioeconomic profile than the native British population (Department of Employment, 1993). Their fertility has declined to match, so that the TFR for immigrant mothers from these groups is about the same level as that of the native population. This is quite different from the experience of the Muslim immigrants from South Asia, from Pakistan and Bangladesh. There the rates of fertility have remained higher and are declining more slowly,

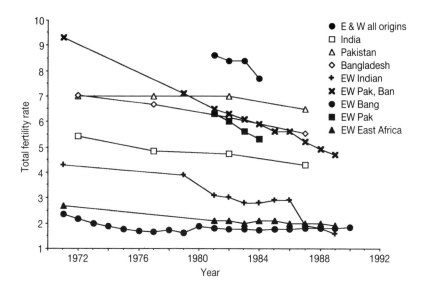

Fig. 7.7. Total fertility rate trends of women born in South Asian countries resident in England and Wales 1971–89, compared with home countries. (Source: Coleman (1994).)

similar to the Maghreb populations but still at a higher level. Bangladesh and Pakistan are both poorer countries than those of the Maghreb and are a long way further behind in their demographic transitions.

This accelerated social change is not only reflected in changes with respect to the timing and number of births, but also in the choice of marriage partner. Populations which in the past had almost invariably married endogamously are presented, upon migration to the West, with a more open society. This presents new and unprecedented opportunities for wider choices of mate, and cultural and intellectual challenges to the traditional systems of social control of mate choice through arranged and/or very early marriage. In all European societies that have received immigrants from outside Europe, novel interethnic unions have arisen. The literature on such marriages is well developed (OECD, 1991; Coleman, 1992) especially in the US (Lieberson & Waters, 1988) and will not be reviewed here. It is worth noting, however, that the incidence of such unions reveals differences in immigrant situations and preferences even more marked than the diverse rates of fertility decline noted above.

Most European data on 'interethnic' marriage are based on nationality data. That would record as exogamous a union formed between a person of foreign nationality and a person of the same foreign origin who had become naturalized, and thereby exaggerate the 'true' rate of interethnic marriage. Marriage data based on birthplace, such as that provided by the French *Enquête Famille* of 1982 (Muñoz-Perez & Tribalat, 1984), can overcome this drawback. Data based on ethnic origin can extend the study irrespective of nationality or birthplace. Almost the only European data of this kind come from the annual Labour Force Survey. This is held in all EC countries but only the British survey asks a question on (self-ascribed) ethnic origin (Jones, 1984). It is a large-scale voluntary sample survey which provides information on the ethnic origin of the partners in current unions. It provides prevalence data, not incidence data. It is customary to pool the data for two or three years to provide a reasonable sample size for the study of ethnic minority marriage. A much more comprehensive source of data on ethnic origins of marriage patterns will be yielded by the 1991 UK Census, but appropriate data were not available at the time of writing.

The Labour Force Survey data yields comparisons such as that shown in Fig. 7.8. The higher propensities for men of immigrant ethnic origin to form unions with indigenous women (except for the Chinese; a consistent finding over several years) is well known from other studies. Note, however, that the difference between men and women varies by ethnic group. There is also considerable variation between groups in the prevalence of interethnic unions. The reasons are complex and are not fully analysed, but appear to

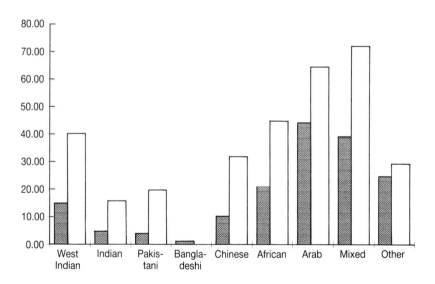

Fig. 7.8. Percentage of men in the UK of different ethnic origins who had white partners, separately for men born abroad (stippled bars) and in the UK (white bars), 1985–88. (Source: Labour Force Surveys 1985–1988.)

follow from a combination of the time of arrival of each immigrant group, its cultural distance from the host population and the extent to which marriage is customarily tightly socially controlled within a limited field of eligibles defined on ethnic, caste or religious grounds (Jones & Shah, 1980). Men of West Indian and African origin are most likely to be involved in unions with white women; all the variables mentioned above are 'favourable' to relatively high levels of interethnic marriage. In general, South Asian rates of interethnic unions are considerably lower, especially among Bangladeshis, the most recent and poorest immigrant group. Lower levels would be expected as South Asian immigrants are not Christian, did not speak English as their first language and come from traditional societies where sexual behaviour and marriage is under strict social control. In Islam nubile girls and married women are (ideally) strictly segregated; in Hindu society marriages are expected to be arranged according to caste norms.

It is the ideal of the immigrant parents to preserve these norms in British society, which many regard as loose and immoral. Many adult immigrants came to the UK, as to other Western countries, with the intention of returning home at a later date. Even though this 'myth of return' may indeed turn out to be a myth for the majority of immigrants, none the less it affects their attitude to integration and to commitment to the host society and to their children's participation in it. However, it is clear that the rules are bending among immigrant populations, more so among Indians (mostly Hindu) than among the Muslim Pakistanis and Bangladeshis. There is little religious opposition to Muslim men marrying non-Muslim women; it is assumed that the wife will convert. It is correspondingly forbidden for a Muslim girl to marry outside the faith, as that automatically becomes a form of apostasy. Muslim immigrant women in Britain (Brown, 1984) and in the rest of Europe are also notably more likely to be illiterate, and have much lower workforce participation rates, than other immigrant women. For all these reasons the proportion of such women in unions with white men is low. The Arab population is somewhat apart from the rest, being a relatively small population of relatively well-off and well-educated men, mostly from Gulf oil states. The sociological literature, reviewed elsewhere, testifies to the strains involved in the social change associated with the new opportunities for interethnic marriage (Benson, 1981).

Time trends, reviewed elsewhere (Tribalat *et al.*, 1991; Coleman, 1992), showed an initial relatively high incidence of interethnic unions when the immigrant populations were still small and dominated by young men; a reduction when immigrant women arrived to redress the sex-ratio imbalance and to stiffen traditional and religious values; and subsequently a slow increase, at least, in Great Britain, among the population of West Indian origin. Interethnic marriage is inhibited by large and socially self-contained immigrant concentrations; in general, interethnic unions are relatively more frequent where the immigrant population size is smaller. The small sample of unions where the ethnic minority partner was born in the UK show a much higher proportion of interethnic unions – about double the porportion compared with men born abroad – and therefore something of a discontinuity between generations in the pace of social change and adjustment. One of the consequences of these trends is the creation of a large (300 000) and growing population of mixed ethnic ancestry, mostly born in the UK. The self-identification of this group, its own choice of partners in unions and its treatment by the 'parental' communities is of the highest importance for the future of integration and assimilation of minorities.

Effects of immigrant populations on size and age-structure of host populations

This leads us to the other major topic of this paper: the effects that immigrant populations have upon the structure of the host population, as opposed to the effects which the host country environment has upon immigrant demographic behaviour. Clearly the potential exists for immigrant streams to transform the size and age-distribution of the total population into which they are migrating. This is a population dynamics question, one relating to the adaptability of populations rather than of their constituent individuals. The answer is to be found in stable population theory, rather than in the sociology or psychology of individual responses.

The question is made sharper by recent interest, both in Europe and in the US, in the relationship between domestic population growth or decline, and population ageing, and the possible importance of immigration streams in counteracting such effects.

Problems of low-fertility populations

Many European populations face the slowdown and end of natural increase, and of the demographic components of labour-force growth, within the next two decades. As we noted above, immigration proceeds at a high level into most European countries despite attempts by governments to limit it. The position is different in the US, where fertility is higher than in Western Europe and immigration, both legal and illegal, has proceeded at a high level for many years. None the less, similar arguments have been made in the US context (Espenshade, 1987b), that the US will in future need even more immigrants in order to ward off population decline and ageing.

Population decline and ageing create two related problems. The first problem arises from an unbalanced, aged population structure. Depending on the birth rate, between 20% and 25% of the population can be expected to be over age 65 in the future, with a correspondingly high aged dependency ratio. The care of this population, both directly in terms of services and indirectly in the form of an adequate workforce to generate the income and tax revenue to pay for its pensions and medical care, may be prejudiced by the small size of the present and likely future population of working age. Immigration, it is claimed, could reinforce the population of working age, provide a service workforce to care for the elderly, reduce the dependency ratio arising out of ageing, and delay or reverse population decline.

The second problem is a potential shortage of workers. The workforce may need immigrant labour, as in the 1960s, irrespective of the increased relative burden of the elderly, because domestic supplies of labour are simply insufficient to match demand. Inadequate labour supply could lead to wage inflation, inefficient use of capital and reduced output. This may apply generally throughout the economy, or more specifically in certain 'dirty' jobs, difficult to mechanize, which the domestic workforce is reluctant to undertake. Finally, the higher average age of the domestic workforce may harm productivity and act as a barrier to innovation, whereas a workforce reinforced by young migrants would be more productive. Economic growth is believed by some to benefit from a growing population, not only to provide the workforce but also to preserve the size of domestic markets and the level of demand and investment confidence for the future (Simon, 1989). This 'mercantilist' approach to the economic and social importance of population growth is, in general, better received across the Atlantic than in Europe.

Can immigration stop population decline or ageing?
In fact a number of theoretical studies and models have shown that international migration can, at a high enough level, stop population decline but cannot eliminate the effects of population ageing arising from low fertility. In the absence of fertility increases, some ageing is inevitable. The ageing process caused by low fertility will eventually stabilize at a constant level if fertility remains constant at any level. But there is an additional component to ageing from further mortality decline in modern populations with very low death rates. Such studies (Coale, 1972; Pollard, 1973, Espenshade, Bouvier & Arthur, 1982; Espenshade, 1987a) have first made it clear that, with sub-replacement fertility, any level of immigration will eventually produce a stationary population at a size (possibly a small size) depending solely on the magnitude of the net immigration. However, the magnitude of immigration required to prevent short- or medium-term

population decline from the current level in a large population such as that of Germany can be very large, depending on the net reproduction rate of the host population. Thus to maintain a stationary population if the TFR of the native population is 1.75 and that of the immigrant population (first generation) is 3.0, an annual immigration of females equivalent to 20% of the annual number of female births would be required (Lesthaeghe, Page & Surkyn, 1988).

Depending somewhat upon the assumptions made about the decline of immigrant fertility rates (given that most immigrants still come from high-fertility countries), high levels of immigration would generate very high levels of population of immigrant origin or descent relatively early next century. The projections of Passel and colleagues, based on the continuation of current US trends (which will forestall population decline, but not ageing, into the next century) indicate that about half the US population would be of non-Hispanic white origin by 2050. Most of the population growth from 1992 (63 million) is expected to come from Hispanic and Asian immigrant flows (Passel & Edmonston, 1992). In the longer run, of course, over two or three hundred years, persistence of high levels of immigration with below-replacement fertility (lower than current US levels) would lead to the progressive replacement of the native-origin population by the immigrant-origin population (Espenshade, 1987a).

Although population decline may be halted by such means, population ageing cannot. Immigrant populations are themselves only a few years younger on average than the host population average and themselves age, requiring further reinforcement to replace themselves. The median age of the immigrants shown in Fig. 7.9 is only 29, compared with 33 for the current US population. It is much more demographically efficient to increase the birth rate. That secures much greater reductions of ageing for the same level of total population change, as has been demonstrated with specific projections and scenarios for the US (Espenshade, 1987a), Italy (Golini, 1993), Belgium (Wattelar & Roumans, 1991); the EC countries (Lesthaeghe *et al.*, 1988), the Netherlands (Kuijsten, 1990) and Germany (Feichtinger & Steinmann, 1992). The last paper showed, subject to the assumptions of the scenarios, that the eventual ratio between the foreign-born and the native-born would depend upon the relative fertility of the two groups and the sex-ratio and age-structure of the immigrants.

Projections of US population sex and age-structure under various assumptions

Some consequences of immigration, in conjunction with various rates of fertility, are presented in a few simple examples below to illustrate the magnitude and tempo of the adaptation of population size and structures to

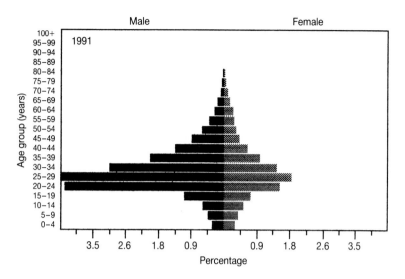

Fig. 7.9. Legal immigrants to the United States, fiscal year 1991, by age and sex. These data are inflated by amnesties granted to former illegal immigrants. (Source: data from US Immigration and Naturalisation Service.)

changes in such external forces. The points are illustrated with four scenarios of US population from 1990 to 2090. The aim of these scenarios is to show the relative demographic effects of fertility change and of immigration upon population size and age-structure. The scenarios share a common starting point with the US population as at the 1990 census, male and female immigration as in fiscal 1991 (851 000 men and 428 000 women), expectation of life for males and females as in 1988 (National Center for Health Statistics, 1991) and the age-specific and total fertility rate as in 1990 (2.08; National Center for Health Statistics, 1993). In each scenario the expectation of life continues to improve at what is hoped to be a plausible level, from the 1988 level of 71.4 years for men and 78.1 years for women to reach 77.6 years for men in 2030 and 83.6 years for women, after which there is no further improvement. No allowance is made for any higher fertility rates among immigrant populations.

In the first scenario, the population proceeds as on current trends in the birth rate, death rate and immigration. In the second scenario, the number of immigrants is increased to double the 1991 gross level. In the third scenario, the immigration level remains at the 1991 level but the birth rate increases (to a TFR of 2.72) by a number initially equivalent to the number of immigrants in 1991, thereby effectively doubling the 'immigration rate' by an increase in births. In the fourth scenario, there is no migration at all but the birth rate is increased so that the population in 2040 is the same as it would have been under scenario 1 under current trends. Basic measurements include total population size, median age, per cent of population aged 65 and over, and the dependency ratio. This is calculated, as is conventional, by dividing the nominally 'dependent' part of the population (0–16 years and 65 years and over) by the nominally 'productive' part, aged 16–64 (per cent). Although these age limits are not wholly realistic in modern conditions, they provide a constant yardstick to measure the general impact of changes in population structure. The 1991 US immigration figures are higher than usual because they incorporate persons who had entered the US illegally but whose presence was regularized in 1991. However, these inflated figures of legal immigration help to compensate for the illegal immigrants entering the US each year who are missing from official data.

Scenario 1: no change

Scenario 1 generates the long-term near-stable population structure of a type with which readers will be familiar (Fig. 7.10). Natural increase eventually ceases around 2050 but population continues to grow at a diminishing 0.1% by 2090 through constant migration. Population size reaches 317 million by 2040, after which it grows much more slowly to reach 328 million by 2090, as expected from demographic theory (Arthur, 1982). The median age increases from about 34 years (female) and 32 years (male) today to a constant 42 years (female) and 40 years (male) by 2040, after which there is no more change. Age composition changes from 22% aged 15 and under in 1990, and 13% aged 65 and over, to reach 18% under 15 and 22% 65 and over, beyond which there is no further significant change. The overall dependency ratio increases from 52 to 68, with the 'aged' dependency ratio doubling from 18 to 36 and the 'youth' dependency ratio falling from 33 to 30, again with little change after 2040 (Fig. 7.11). These are quite modest changes and burdens compared with some forecast for European populations of lower fertility.

Scenario 2: more immigrants

In the second scenario, where the immigration rate is increased to 3.6 million and everything else left as in simulation 1, population grows considerably faster, to 343 million in 2040 and to 376 million in 2090, although the rate of growth diminishes substantially after 2040. The consequences for median age are relatively modest. Median age of females stabilizes at 42 and of males at 41. Overall dependency ratio stabilizes at 66, 36 for the aged dependency ratio and 30 for the youth dependency ratio.

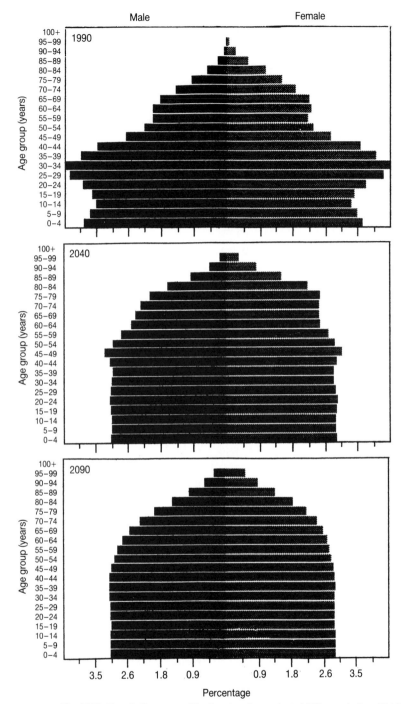

Fig. 7.10. Population pyramids showing the projected US population (%) by age and sex in 1990 (base year) 2040 and 2090 according to Scenario 1; see text for explanation. (Source: projections calculated for this chapter.)

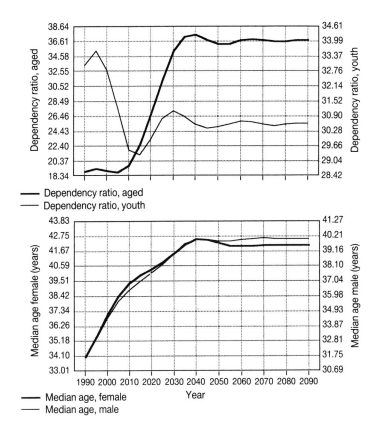

Fig. 7.11. Population dependency ratios and median age by sex, for projected US population 1990 (base) to 2090, according to Scenario 1; See text for explanation. Aged dependency ratio = population 65 and over/population 15–64. Youth dependency ratio = Population under 15/population 15–64. (Source: projections calculated for this chapter.)

Scenario 3: *higher fertility*

In the third scenario, the annual births are increased by a number equivalent to the number of immigrants at the 1991 level. That increases the TFR from 2.1 to 2.72. The consequent population growth is spectacular, rising to 334 million by 2040 and 407 million by 2090, and continues to increase exponentially. Natural increase remains around 1% per year throughout and this approximates to the population growth rate in the long run, as immigration (being fixed in volume) becomes proportionately less in comparison with population size, and the population structure accordingly approximates more and more closely to a stable distribution. Uniquely in this scenario, the population aged under age 15 actually increases from 22% to 25% of the total. Even then, ageing cannot be stopped but the proportion of the elderly aged 65 and over increases in this simulation only from 13% to 15% of the population. Median age falls slightly after an inevitable increase early next century (from the baby-boom cohorts) stabilizing at 33 years for women and 32 years for men. This creates an overall dependency ratio little different from other simulations (the smallest dependency ratios are always those of non-growing populations, irrespective of their form). But its composition is quite different. The aged dependency ratio increases from 18 to a constant 25; the youth dependency ratio increases from 33 to 42.

Scenario 4: *fertility replaced immigration*

Finally, the last simulation has a TFR fixed at a level such that the population size attained by 2040, without immigration, is the same as that attained in scenario 1 with immigration on the level of 1991 and a somewhat lower TFR. The level of population by 2040 which was produced in scenario 1 with over 1 million net migrants per year and a TFR of 2.1 is also achieved in scenario 4, in the absence of immigration, with an increase of fertility to a TFR of 2.185. That is equivalent to every tenth family having an additional child. As intended, population growth is similar to that in simulation 1, ending in 2090 at 343 million but, unlike scenario 1, growing at a modest exponential rate of 0.2% per year. In the absence of migration a completely stable population distribution will be approached. The population aged under 15 falls from 22% to a constant 20%; the population aged over 64 increases from 13% to 20%. The overall dependency ratio is little different from that in the other simulations and stabilizes at 68. The aged ratio finalizes at 34 and the youth ratio at 33. Median age becomes 40 for women and 38 for men.

The results of the simulations are summarized in Table 7.3. They illustrate in detail a number of points already generally known from demographic theory. Other things being equal, fertility has much more potent effects on populations than immigration. No plausible level of immigration is able to stop population ageing, but a return to the fertility levels that were commonplace in the 1960s will do so. The 1991 levels of immigration are equivalent, in population growth terms up to 2040, to an increase in fertility of just 5%. Furthermore, the attainment of the same population size through the route of slightly enhanced fertility reduces median age and the aged dependency ratio compared with the same population growth attained through constant migration. On the other hand, substantial increases in migration over the current level (scenario 2), although

Table 7.3. *Summary table, US population scenarios* 1990–2090

	Scenario 1 Current trends	Scenario 2 Migration doubled from 1991 gross level	Scenario 3 TFR rises to 2.72 migration as at 1991	Scenario 4 TFR rises to 2.18 no migration
TFR 1990	2.08	2.08	2.08	2.08
TFR 1995	2.08	2.08	2.73	2.18
e0m 1990	71.44	71.44	71.44	71.44
e0f 1990	78.07	78.07	78.07	78.07
e0m 2030	77.63	77.63	77.63	77.63
e0f 2030	83.61	83.61	83.61	83.61
migration m	851	2430	851	0
migration f	428	1222	428	0
pop 1990	249	249	249	249
pop 2010	282	293	303	281
pop 2020	297	313	334	295
pop 2040	317	343	407	317
median age 1990	33	33	33	33
median age 2030	40	40	33	39
median age 2060	41	41	33	39
% aged ≥65 1990	13	13	13	13
% aged ≥65 2030	21	21	18	21
% aged ≥65 2060	22	22	15	21
DR ≥65 1900	19	19	19	19
DR ≥65 2030	35	34	31	35
DR ≥65 2060	37	36	25	34

Note TFR, total fertility rate; e0, expectation of life at birth; migration in thousands per year; population in millions; DR ≤ 65, aged dependency ratio (population over age 64/population aged 15–64) × 100

increasing population by 26 million by 2040, secured no significant reduction in the median age or in the proportion aged over 65 or in the aged dependency ratio. It would, however, considerably enhance the fraction of the population of immigrant origin in the future, above that projected by Passel & Edmonston (1992).

Conclusions
These simple empirical results, from real and from simulated populations, have attempted to show the pace and impact of migration on population behaviour at the level of the individual and at the level of the population

itself. Immigrant populations show a tendency to adapt their behaviour, most likely irreversibly, to new cultural and economic pressures. The response of individuals in terms of social change, as measured through their population averages, is highly varied according to the nature of the society of origin and, probably, to the interaction of the immigrants with the host society. We do not yet have an adequate predictive model for forecasting the fertility or intermarriage response of immigrant populations that have put themselves into new environments where their traditional demographic regime is challenged. The range of that response, from 'defensive structuring' to accelerated demographic transition, is very great and, of course, has substantial consequences for future population composition. No formal analyses of the variability of this response have been attempted, as there are more confounding variables than there are data. But the pattern of demographic response, at least in respect of fertility, generally reflects in an accelerated form the demographic transition occurring in the country of origin. However, the pace of change varies considerably between immigrant groups. The impact of the different environments of the various Western countries themselves, into which these populations have settled, seems much less potent. While individuals clearly change their attitudes when confronted with new realities, the change in behaviour between successive generations of immigrant populations, especially between overseas- and local-born, is much more marked. In immigrant populations, the plastic characteristics of such adaptive behaviour are much more marked between generations than within them

The demographic response at the population level contains some elements, derived from fundamental demographic theory, which may seem counterintuitive, for example the inability of immigration to counteract population ageing, and the certain acquisition, in populations with fertility below the replacement level, of a stationary population size at any level of positive net immigration. At the population level, the response of demographic parameters (ageing, median age, size and so on) is of course measurable and predictable. Populations are plastic in that they forget their past and acquire a size and shape moulded by the pressures of current and recent vital rates and of immigration. But their inertia is considerable and changes are not reasonably stable until the lapse of about two generations. The new shape then becomes permanent for as long as the new regime of vital rates and immigration persists. Demographers, it seems, have been using concepts of 'plasticity' for decades, without realizing it.

References
Arthur, W. B. (1982). The ergodic theorems of demography: a simple proof. *Demography* **19**, 439–45.

Bean, F. D. & Frisbie, W. P. (1978). *The Demography of Racial and Ethnic Groups.* New York: Academic Press.

Benson, S. (1981). *Ambiguous Ethnicity: Inter-Racial Families in London.* Cambridge University Press.

Bongaarts, J. & Potter, R. G. (1983). *Fertility, Biology and Behavior: an Analysis of the Proximate Determinants.* New York: Academic Press.

Brown, C. (1984). *Black and White Britain.* London: Policy Studies Institute/ Heinemann Educational.

Chesnais, J.-C. (1993). *The Demographic Transition,* transl. E. Kreager & P. Kreager. Oxford: Clarendon Press.

Coale, A. J. (1972). Alternative paths to a stationary population. In: *Demographic and Social Aspects of Population Growth,* ed. C. F. Westoff, R. Parke, *et al.,* United States Commission on Population Growth and the American Future. Washington D.C.: USGPO.

Coleman, D. A. (1992). Ethnic intermarriage. In: *Minority Populations: Genetics, Demography and Health. Proceedings of the twenty-seventh annual symposium of the Galton Institute, London, 1990,* ed. A. H. Bittles & D. F. Roberts, pp. 208–40. London: Macmillan.

Coleman, D. A. (1994). Trends in fertility and intermarriage among immigrant populations in Western Europe as measures of integration. *Journal of Biosocial Science* **26**(1), 107–36.

Coleman, D. A. (1995). Fertility in Britain - variation and its significance. In: *Human Populations,* ed. V. Reynolds & A. J. Boyce, Oxford University Press (in press).

Department of Employment (1993). Ethnic Origins and the Labour Market. *Employment Gazette* **101**(2), 25–43.

Espenshade, T. J. (1987a). Population dynamics with immigration and low fertility. In: *Below-Replacement Fertility in Industrial Countries,* ed. K. Davis, M. S. Bernstam & R. Ricardo-Campbell, pp. 248–61. New York: The Population Council. (Supplement to Vol. 12, Population and Development Review.)

Espenshade, T. J. (1987b). *Why the United States Needs Immigrants.* Washington: The Urban Institute.

Espenshade, T. J., Bouvier, L. F. & Arthur, W. B. (1982). Immigration and the stable population model. *Demography* **19**, 125–33.

Feichtinger, G. & Steinemann, G. (1992). Immigration into a population with fertility below replacement level–the case of Germany. *Population Studies* **46**(2), 275–84.

Frisch, R. E. (1978). Population, food intake and fertility. *Science* **199**, 22–30.

Golini, A. (1993). La tendenze demografiche dell'Italia in un quadro europeo. In: *Tendenze demografiche e politiche della popolazione.* Rome, 19 March 1993: Consiglio Nazionalie delle Ricerche: Istituto de Ricerche sulla Popolazione.

Isnard, E. (1992). La fécondité des étrangères en France se rapproche de celle des Françaises. *INSEE Première,* No. 231, Novembre 1992.

Jones, P. R. (1984). Ethnic inter-marriage in Britain: a further assessment. *Ethnic and Racial Studies* **7**, 398–405.

Jones, P. R. & Shah, S. (1980). Arranged marriages: a sample survey of the Asian case. *New Community* **8**, 339–43.

Keyfitz, N. (1977). *Introduction to the Mathematics of Population.* New York: Addison-Wesley.

Kuijsten, A. (1990). The impact of migration streams on the size and structure of the Dutch population. Paper presented at *The Demographic Impact of International Migration*, a conference held at NIAS, Wassenaar, in September 1990.

Lasker, G. W. (1985). *Surnames and Genetic Structure*. Cambridge University Press.

Lasker, G. W., Mascie-Taylor, C. G. N. & Coleman, D. A. (1986). Repeating pairs of surnames in marriages in Reading (England) and their significance for population structure. *Human Biology* **58**(3), 421–5.

Leridon, H. & Menken, J. (1982). *Natural Fertility: Patterns and Determinants*. Liège: Editions Ordina.

Lesthaeghe, R., Page, H. & Surkyn, J. (1988). Are immigrants substitutes for births? In: *IDP Working Paper 1988-3*. Brussels: Free University Brussels, InterUniversity Programme in Demography.

Lieberson, S. & Waters, M. C. (1988). *From Many Strands: Ethnic and Racial Groups in Contemporary America*. New York: Russel Sage Foundation.

Livi-Bacci, M. (1991). *Population and Nutrition. An Essay on European Demographic History*. Cambridge University Press.

Lunn, P. G. (1988). Malnutrition and fertility. In: *Natural Human Fertility, Social and Biological Determinants*, ed. P. Diggory, M. Potts & S. Teper, pp. 135–52. London: Macmillan.

McFarland, D. D. (1969). On the theory of stable populations: a new and elementary proof of the theories under weaker assumptions. *Demography* **6**, 301–22.

Menken, J., Trussell, J. & Watkins, W. (1981). The nutrition fertility link: an evaluation of the evidence. *Journal of Interdisciplinary History* **11**, 425–41.

Muñoz-Perez, F. & Tribalat, M. (1984). Mariages d'étrangers et mariages mixtes en France: évolution depuis la Première Guerre Mondiale. *Population* **39**(3), 427–62.

National Center for Health Statistics (1991). *Vital Statistics of the United States 1988*, Vol. II, Section 6, *Life Tables*. Hyattsville, Maryland: US Department of Health and Human Services, Public Health Service.

National Center for Health Statistics (1993). *Monthly Vital Statistics Report*, Vol. 41, no. 9 (supplement), *Advance Report of Final Natality Statistics 1990*. Hyattsville, Maryland: US Dept of Health and Human Services, Public Health Service.

Ni Bhrolchain, M. (1993). Recent fertility differentials in Britain. In: *New Perspectives on Fertility in Britain*, ed. M. Ni Bhrolchain (*Studies on Medical and Population Subjects* No. 55), pp. 95–109. London: HMSO.

OECD (1991). *Migration: The Demographic Aspects*. Paris: OECD.

Passel, J. S. & Edmonston, B. (1992). *Immigration and Race: Recent Trends in Immigration to the United States*. Washington, DC: The Urban Institute.

Pollard, J. H. (1973). *Mathematical Models for the Growth of Human Populations*. New York: Cambridge University Press.

Poulain, M. (1993). Confrontation des statistiques de migrations intra-Européennes: vers plus d'harmonisation? *European Journal of Population* **9**(4), 353–81.

Robinson, V. (1986). *Transients, Settlers and Refugees: Asians in Britain*. Oxford: Clarendon Press.

Robinson, V. (1988). The new Indian middle class in Britain. *Ethnic and Racial Studies* **11**, 456–73.

Siegel, B. J. (1970). Defensive structuring and environmental stress. *American*

Journal of Sociology **76**, 11–32.

Simon, J. L. (1989). *The Economic Consequences of Immigration*. Cambridge: Basil Blackwell.

Tribalat, M., Garson, J.-P., Moulier-Boutang, Y. & Silberman, R. (1991). *Cent Ans d'Immigration, Étrangers d'Hier, Français d'Aujoud'hui*. Paris: Institut National d'Études Démographiques / Presses Universitaires de Paris. (Travaux et Documents de l'INED, Cahier no. 131).

Wattelaar, C. & Roumans, G. (1991). Simulations of demographic objectives and migration. In: *Migration: Demographic Aspects* (ed. OECD), pp. 57–67. Paris: OECD.

8 The use of surnames in the study of human variation and plasticity

JOHN H. RELETHFORD

Summary

A central idea in studies of human adaptation is that human beings can adjust to changing environmental conditions through a variety of mechanisms. Although debate continues regarding specific definitions and a general classification of modes of adaptation, most authors consider that human adaptation occurs via four broad processes: genetic evolution (natural selection, mutation, etc.), developmental plasticity, acclimatization, and cultural patterns of behavior (Frisancho, 1979). Other chapters in this book review these concepts, discuss alternative methods of study design and analysis, and provide case studies of human adaptation. This chapter looks at how surnames, a trait not affected by adaptive processes, can none the less be useful in studies of adaptation and plasticity. Surnames can provide information on genetic relationships between individuals and populations. This information can be useful in studies of adaptation by: (1) providing the expected patterns of variations under a neutral model (no adaptation); (2) providing estimates of demographic and cultural factors of interest to studies of variation and plasticity, especially mate choice and migration; and (3) providing estimates of genetic relationship between individuals for use in studies of genetic epidemiology.

Introduction

Studies of human variability and plasticity generally focus attention on traits of specific interest to hypotheses of plasticity and adaptive change. Relationships are examined between these traits and suggested adaptive or selective forces. Some classic examples include hemoglobin variants and malaria (Livingstone, 1958), body size and shape and climate (Roberts, 1978), skin color and ultraviolet radiation (Robins, 1991), and chest morphology and high-altitude environments (Frisancho, 1979). A key focus here is the identification and analysis of traits suspected to have significance, through the process of natural selection (genetic adaptation) and/or through developmental change (plasticity).

146

Another tradition in human biology involves the selection of traits that are presumed to have no adaptive significance; that is, they are neutral. These adaptively neutral traits are used collectively to construct measures of biological distance between human populations, with the goal of interpreting these distances in terms of population history and demographic and cultural influences on migration, genetic drift, and mating patterns.

Of course, it is often difficult to divide traits into those that have adaptive significance and those that do not. Blood groups and other genetic markers were first hailed as completely neutral, and therefore most appropriate to studies of population history. Many of these genetic markers, however, appear to have been affected by natural selection (Livingstone, 1980). Anthropometric traits have been used in studies of both adaptation and population history. Cranial measures are often interpreted as evidence of climatic adaptation when comparing groups across large geographic distances such as continents (Roberts, 1978; but see Howells, 1973, 1989). Over shorter distances, these same measures often provide valuable insight into population history (Relethford, 1988a; Brace & Hunt, 1990). Another distinction between adaptation studies and population history studies is their consideration of covariates. Age, for example, is a key variable in studies of growth and plasticity. Often the relationship of a trait with age is of prime importance. In studies of population history, age is often 'noise' that must be adjusted for statistically in order to facilitate analysis.

One trait that has been used widely in studies of population history, which is completely neutral in terms of environmental effects, is surnames. In a number of societies a father's last name is passed on to his sons. This mode of cultural transmission is analogous to genetic transmission of a Y-linked gene. This similarity of cultural and genetic inheritance has allowed a number of methods to be developed that provide insight into microevolutionary processes. To date, surname frequencies have been used to examine a wide variety of problems and processes in human biology, including inbreeding, marriage preferences, genetic drift, migration and gene flow, admixture, and genetic epidemiology. The use of surnames as neutral genetic traits has also allowed genetic inferences to be made from historical data. We can thus look at the evolutionary dynamics of populations from decades or centuries in the past, well before the discovery of genetic markers.

Several recent reviews have documented the usefulness of surnames in studies of population history and genetic structure (Lasker, 1985; Relethford, 1988b). Although surnames cannot be used for all human groups, and although there are often problems in analysis and interpretation, they have proven successful in many cases. What use, however, do surnames have for studies of human variation and plasticity? At first glance, none. After all, surnames are biologically neutral. Although surnames do change over time (name changes and other forms of 'mutation') and are often based on factors such as occupation (e.g. 'Smith'), these changes are cultural in nature. They do not reflect the type of environmental interaction looked at in adaptation studies. Surnames are not affected by climate or diet or altitude. Considering their environmentally neutral nature, surnames appear to have absolutely no relevance to the study of human adaptation.

However, it is precisely their neutrality that makes surnames a valuable supplement to studies of adaptation. This point has been recognized by Gabriel Lasker, who was a central figure in the emergence of studies of both adaptation and surnames. The purpose of this chapter is to outline three ways in which surnames can be useful in human adaptation research. First, surnames can be used to provide estimates of the expected pattern of *neutral* variation among human populations that provides a null model against which to compare other traits. Second, surnames can provide valuable information on patterns of migration, mating preferences, inbreeding, and other demographic and cultural variables that might be useful in interpreting patterns of human adaptation. Because records of surnames have been around much longer than genetic markers, they also have the advantage of being useful in historical studies. Third, surnames can be used as proxy measures of genetic relationship in epidemiologic studies. Again, the fact that surname records have been around for a long time provides such measures in cases where genetic and/or genealogical data are not available. Several case studies will be presented to illustrate these three approaches. It must be emphasized that surname analysis will not always be useful in any given study of human adaptation. Often, other data might be available that serve these functions more appropriately. Also, there are certain assumptions of surname analysis that might render it of limited value in certain cases. Finally, surnames do not always follow the paternal mode of inheritance. In any case, the main point here is that under certain conditions, surname analysis can provide a useful *supplement* in the study of human adaptation and plasticity.

Surname methods

Before elaborating on the points discussed above, it is useful to review briefly some of the methods and limitations of surname analysis. This section is intended only as a brief review, and the reader is encouraged to seek out the appropriate references for additional information, particularly on specific mathematical formulae.

Although there had been a few earlier studies linking surname distribution to genetic relationships (see Lasker, 1985), the major impetus to the development of surname genetics came with the 1965 publication of Crow & Mange's now classic paper. As noted above, many cultures (especially Western societies) inherit the surname from the father. Crow & Mange (1965) noted that there is a relationship between consanguinity and the probability of a marriage being *isonymous*: that is, one where both husband and wife have the same last name (as always, reference to the wife's surname indicates her *maiden* name). For example, the probability of two first cousins of opposite sex having the same surname is 1/4 and the inbreeding coefficient of their children is 1/16. Likewise, the probability of two second cousins having the same surname is 1/16 and the inbreeding coefficient of their children is 1/64. In general, the probability of an isonymous marriage is four times the inbreeding coefficient. Therefore, one could estimate the average inbreeding coefficient in a population by computing the proportion of isonymous marriages and dividing this number by 4. In general terms, $P = 4F$, where P is the proportion of isonymous marriages and F is the inbreeding coefficient. Not all possible genealogical relationships satisfy this condition, but it does hold true for the more common relationships encountered in the human populations for which surnames are available (Crow & Mange, 1965; Crow, 1980, 1989). Extensions are also available for societies in which children inherit *both* their father's and mother's last name (e.g. Pinto-Cisternas, Pineda & Barrai, 1985).

Crow & Mange's method makes a number of assumptions (Crow, 1980, 1983; Lasker, 1985). One of the most serious of these is the assumption that common surnames reflect common ancestry, and not some other form of common origin. An example of a violation of this assumption would be where two individuals have the same surname because of ancestors practicing the same occupation: in many cases, surnames originated from occupation, such as 'Smith'.

How accurate is the surname method for estimating inbreeding? In many case studies the estimates of inbreeding based on surnames are higher than those obtained directly from genealogies. This difference has often been interpreted as due to multiple surname origins inflating the surname estimate and incomplete ascertainment of remote ancestors deflating the genealogy estimate. Laurine Rogers (1987) examined concordance of inbreeding estimates and suggested that differences in sample composition (different samples for surnames and genealogies) might also influence this difference. Another problem with the surname method is that the inbreeding estimate is relative to a hypothetical founding population where each person has a unique surname (Crow, 1983; Alan Rogers, 1991). Since such a hypothetical case is unlikely to ever occur in a real population, comparison of inbreeding coefficients from surnames is somewhat limited.

From an anthropological perspective, Crow & Mange's method is more useful for assessing patterns of mate choice than estimating inbreeding. They noted that by looking at all possible combinations of male and female names one can derive the estimated degree of inbreeding that is expected under the assumption of random mating (that is, each male as equally likely to marry any given female). By comparing observed inbreeding with the random expectation, one can further derive an estimate of non-random inbreeeding. For example, if total inbreeding is significantly greater than random inbreeding, then a pattern of preferential consanguineous marriage is suggested. On the other hand, if total inbreeding is less than random inbreeding, then avoidance of marriage between relatives is suggested. The partitioning of total inbreeding into random and non-random components is therefore potentially useful in assessing the overall effect of behaviors affecting mate choice.

A problem of Crow & Mange's method is that in most populations the actual number of isonymous marriages is usually low. In one study, for example, only 72 out of 5791 marriages (1.2%) were isonymous (Relethford & Jaquish, 1988). The usual small number of such marriages means that estimated parameters frequently have large standard errors (Lasker, 1985). An alternative method, developed by Lasker & Kaplan (1985), gets around this problem by considering the frequency of shared *combinations* of surnames, called repeating pairs. For example, consider a marriage between a man with the last name of Jones and a woman with the last name of Smith. Lasker & Kaplan's repeating pairs (RP) method looks at the proportion of all marriages that are between Jones and Smith. Thus, information on *all* marriages is important in deriving the RP statistic. Chakraborty (1985) extended their method to allow analytic computation of the value of RP expected at random. By comparing observed and expected values of RP, a researcher can make inferences regarding factors influencing mate choice in the same manner as the Crow & Mange method, but with greater statistical power.

Surnames can also be used to provide estimates of genetic similarity between individuals within populations and among populations. Here, marriage data are not needed, but only lists of surnames. All of these methods essentially match all possible individuals by surname to compute an estimate of expected genetic similarity. By looking at such estimates between pairs of populations, one can make statements regarding relative genetic similarity, such as population A and population B being more similar than either is to population C. The goal of such work is to determine the reasons why some groups are more similar to each other. These reasons can include geographic proximity, ethnic affiliations, cultural barriers, and population history, among others. Newton Morton and colleagues (1968) were perhaps the first to use surname frequencies in this manner as part of their efforts to estimate 'genetic kinship'. Investigation of population affinities from surnames did not reach a wide anthropological audience, however, until Lasker (1977) presented his own independently derived 'coefficient of relationship'. To date, Lasker's method has been the most widely used measure of population similarity used in surname studies. Recently, Relethford (1988b) derived similar estimates of genetic kinship and distance for use in studies of population history and structure. These measures, along with other variants, are all similar mathematically and provide a convenient measure of the pattern of among-group variation.

Surnames as measures of neutral variation

One potential use of surnames in studies of human variation and plasticity comes from the fact that surnames are neutral with respect to environmental influences. Here, surnames are used to construct proxy measures of genetic similarity that can be used as null hypotheses of neutral variation. Other observed patterns of phenotypic variation, based on measures such as anthropometry or physiology, can then be compared to the null model. This philosophy is best expressed by Lasker, who notes that

> Surname models . . . isolate genetic aspects and deal with them separately. The more one believes genetic-environmental interaction to be important in human biology, the more reason there is to start one's analysis in situations in which one or the other factor is minimized. (Lasker, 1985, p. 3)

An example of surnames and morphological plasticity

In my own work on nineteenth century anthropometric data from western Ireland, I found a significant degree of variation in measures of body size among adult males in five small populations (based on stature, hand length, and forearm length) (Relethford, 1988a). Two other populations were studied, but excluded here due to extensive English admixture. Some populations were taller and with longer arms than others. Why? One possibility might be population differences in nutrition or some other environmental factor affecting adult body size. Of course, it is also possible that this among-group variation is a reflection of genetic structure; that is, due to genetic drift and gene flow. One way of testing this is to look at the pattern of among-group variation observed from surname frequencies, which are neutral and which would not reflect any environmental differences. I computed a matrix of among-group distances based on the surname frequencies of 1530 families from the five populations, using methods outlined elsewhere (Relethford, 1988b). I then compared the distance measures based on body size and surname frequencies. The correlation between these two distance measures is 0.854, which is highly significant according to the Mantel test ($p < 0.001$). On the basis of this large and significant correlation, it is reasonable to conclude that among-group variation in body size reflects an interaction of genetic drift and gene flow, and not the impact of environmental factors producing developmental plasticity.

In spite of the potential utility of such comparisons, this method has not been widely employed in studies of human adaptation. More often, estimates of population similarity and distance from surnames are compared to other estimates based on traits presumed to also be neutral. The goal here has been comparison of different types of data in order to see if the expected concordance among traits due to common population history and structure actually applies. Although such research can in theory lead to suggestive hypotheses regarding adaptation (such as traits that do not fit an expected pattern), there has been no real effort to build in a hypothesis of adaptation from the beginning. In other words, there has not been much interaction between the two major types of human biology studies discussed above: adaptation and population history and/or structure. Surnames could potentially be useful in examining neutral genetic relationships among groups, as are genetic markers. Surnames have certain advantages here: they are cheap and easy to collect, and they can be collected from many populations (past and present) for which genetic marker data are not available.

Surnames as measures of demographic variability and plasticity

Any study of human adaptation must at some point consider a variety of demographic and cultural variables. Some of the more important variables include mating preferences and level of inbreeding, migration patterns and rates, and changes in population size. Thus, human variability and plasticity can be considered at the population level. Assessment of migration is particularly important in adaptation studies: comparison of migrants and sedentes is a widely used and successful strategy (Mascie-Taylor & Lasker, 1988). All of these variables can be of interest either at the level of the individual case or at the level of a sample. For the latter, data are often available for direct assessment of cultural variation. Such data can include census data, genealogies, interviews, and other measures.

In many cases, however, direct data on marital structure and migration are not readily available. In such cases, surnames can be used to provide relevant information. As noted above, both the Crow–Mange method and the repeating pairs method of surname analysis can provide information on average patterns of mate choice and inbreeding. Also, estimates of random inbreeding and the random component of the RP measure often show high correlations with known rates of migration as expected from population genetics theory (Relethford, 1992). Thus, surnames can be used as surrogate measures of migration rates, providing at the very least estimates of relative gene flow. Lacking any other data, surnames might be useful in identifying which samples have experienced greater levels of immigration and which are more endogamous.

One use of such methods has been in studies examining potential biological effects of inbreeding. For example, Lasker & Kaplan (1974) looked at a number of anthropometric measures from both Mexico and Peru in an attempt to determine if inbreeding had any significant effect on phenotypic variance (inbreeding depression). Both samples were divided into subsamples defined on the basis of whether their parents has the same surname. Lasker & Kaplan concluded that there was little evidence to support an effect of inbreeding depression for most anthropometric traits. Their basic methodology might be of use in future studies of human adaptation. There is some continued debate over causes of generational changes in human morphology: do they reflect adaptive changes related to environmental changes, or do they reflect phenotypic changes expected to accompany breakdown of isolation (heterosis)? In cases where genealogical data are lacking, surname analysis can be of use in testing the null hypothesis of heterosis (given, of course, adequate numbers of isonymous marriages).

Surnames can also be used to look at the possible relationship of inbreeding and prevalence of recessive genetic disorders. Jorde & Morgan (1987) found low levels of inbreeding among the Utah Mormons based on surname analysis, which is in agreement with pedigree analysis. Although not conclusive, such low levels of inbreeding may relate to the low levels of birth defects in Mormons (Jorde, 1982) and low levels of hemochromatosis, a recessive disease (Jorde & Morgan, 1987).

Another example of the relationship of disease prevalence and inbreeding estimated from surnames is Sorg's (1983) study of diabetes in Vinalhaven, Maine. She found low levels of inbreeding on the island. This result rejects an earlier suggestion of a link between high inbreeding and diabetes prevalence on the island. Sorg points out that the results although they do not rule out a genetic component, suggest that non-genetic factors (such as age, diet, and obesity) are more important in diabetes prevalence. This study provides an additional example of how surnames can be used to test the null hypothesis of genetic factors at a population level.

Surnames in epidemiological analysis

Some of the above studies focus on indirect use of surname measures. Here, inferences are made based on statistics obtained from entire *samples*, such as disease rates or levels of inbreeding. Surnames can also be used to focus attention on possible genetic factors in epidemiologic research at the *individual* level. In this context, surnames are used instead of more traditional (but sometimes lacking) data on genetic relationships obtained from genealogies and/or genetic markers.

Furthermore, by using surnames to assess the role of genetic factors in epidemiology, researchers may be better able to understand nongenetic influences, such as developmental plasticity. As in all studies of human variation, the basic question in epidemiology is: to what extent are differences in disease susceptibility (and other epidemiologic measures) related to genetic factors, the physical environment, the sociocultural environment, or developmental plasticity? In the absence of detailed data on genetic markers and genealogical data, surnames can provide information needed for such distinctions.

Admixture studies are one area of potential use. Comparison of samples with different levels of admixture can often be useful in assessment of relative genetic and environmental influences on disease. Some examples include investigation of diabetes in Native American and Mexican American populations, and hypertension in African American populations (Chakraborty, 1986). Individual ancestry is often assigned based on self reports and/or evidence from genetic markers. In some cases, surnames could also be used to classify individuals by degree of admixture. Chakraborty *et al.* (1989) investigated admixture classification methods using samples from northern Chile and western Bolivia. They compared results from genetic markers and their own method of surname analysis. They found that surname analysis provided a reasonable correspondence with results from genetic marker analysis. Thus, surnames can be used as a cost-effective means of measuring admixture under appropriate conditions. Their method has great potential for other genetic epidemiologic studies of human admixture and disease.

Surnames can also be used as a proxy measure of genetic relationship in case-control studies. Individuals paired on the basis of surname are expected to be, on average, more similar genetically than any random pair of individuals. This type of approach has been used by Swedlund, Meindl & Gradie (1980) in their investigation of longevity in historical Massachusetts. Using vital records and other genealogical data, they draw pairs of individuals at random and pairs at random but with the same surname. In addition, they formed pairs composed of sibs. For each of these three samples (sib pairs, surname pairs, random pairs) they computed correlation coefficients for age of death for both members of each pair. The sib-pair correlations were significant and the random-pair correlations were not significant. Thus, age of death is more similar between sibs than between any two randomly chosen individuals. This result supports the hypothesis that longevity is at least partly influenced by genetic factors.

From the perspective of surname analysis, the interesting part of their study is that the surname-pair correlations were also significant, although not as large in magnitude as the sib-pair correlations. Thus, cases defined purely on the basis of identical surnames gave the same general conclusion (suggestion of a genetic component in longevity) as that obtained from sib-pairs. This finding means that, in the absence of genealogical data, surnames can be used to construct samples of presumably related individuals with which to compare samples of unrelated individuals. Of course, identical surnames do not guarantee genetic relationship (perhaps why the surname-pair correlations were lower in magnitude in Swedlund *et al.*'s study). Data on actual genealogical relationships are better suited to studies of genetic epidemiology. The main point here is that such data are not always available, whereas surnames are often readily accessible. The study of Swedlund *et al.* provides reassurance that surnames can be used as proxy measures of genetic relationship.

In another study, Cleek (1989) used parental isonymy to investigate possible genetic factors in cancer mortality. Using data from Wisconsin vital records for 1979 to 1985, he looked at the frequency of parental isonymy among cancer deaths and compared it with the frequency expected at random (obtained from controls, defined as non-cancer deaths above the age of 60). There were significantly more isonymous female leukemia cases than expected, consistent with an expectation of partial genetic influence. Another significant comparison was between male lung cancer deaths and male controls. Here, there were significantly fewer cases of isonymy among cases than among controls. This finding fits the idea that there are tumor-suppressing genes for lung cancer. Cleek also looked at different combinations of parents' names (not necessarily isonymous) among cancer deaths and controls, and found similar results.

Cleek (1989) also looked at the frequency distributions of individual surnames among cancer deaths. He found that for almost every type of cancer there were several specific surnames that showed significant differences when compared to control samples. Cleek also performed cluster analysis on cancer sites and surnames, finding suggestive evidence of genetic influence on a variety of cancers, especially breast, lung, and colon cancer. Cleek's study is valuable in showing the variety of ways in which surnames can be used to detect genetic relationships with disease.

Conclusions

The use of surnames in genetic analysis dates back to early attempts in the nineteenth century (Lasker, 1985), although almost all studies came after, and were strongly influenced by, Crow & Mange's (1965) paper. Since that time, many studies of surname methods and applications have been published. Most of these studies focus primarily on genetic structure, particularly the investigation of inbreeding, mate choice, and migration. Surname analysis is not, and has not been, a main method used in studies of human adaptation and plasticity, even though Gabriel Lasker has long championed the use of surnames in a variety of studies of human variation.

This chapter has reviewed some of the ways in which surname analysis might benefit the study of human adaptation. Surnames provide inexpensive and easy-to-collect data on neutral genetic relationships within and among human populations. Such data can provide valuable information regarding genetic relationships as well as demographic patterns. In particular, studies such as those performed by Chakraborty *et al* (1989), Swedlund *et al.* (1980) and Cleek (1989) illustrate how surnames can be incorporated into epidemiologic study designs.

Neutral genetic models, patterns of mate selection, migration, and genetic relationships among individuals and groups are all factors that must be taken into account in analyses of human variation and plasticity. It is important to stress that surnames can provide information on all of these factors. Surnames are particularly useful when genealogical and direct genetic data are not available. In such cases, surnames, which are very often easily available, can allow application to the study of human variation and plasticity.

Acknowledgments

I thank Barry Bogin and Gabriel Lasker for their helpful comments and suggestions.

References

Brace, C. L. & Hunt, K. D. (1990). A nonracial craniofacial perspective on human variation: A(ustralia) to Z(uni). *American Journal of Physical Anthropology* **82**, 341–60.

Chakraborty, R. (1985). A note on the calculation of random RP and its sampling variance. *Human Biology* **57**, 713–17 (and errata in **58**, 991).

Chakraborty, R. (1986). Gene admixture in human populations: Models and predictions. *Yearbook of Physical Anthropology* **29**, 1–43.

Chakraborty, R., Barton, S. A., Ferrell, R. E. & Schull, W. J. (1989). Ethnicity determination by names among the Aymara of Chile and Bolivia. *Human Biology* **61**, 159–77.

Cleek, R. K. (1989). Surnames and cancer genes. *Human Biology* **61**, 195–211.

Crow, J. F. (1980). The estimation of inbreeding from isonymy. *Human Biology* **52**, 1–14.

Crow, J. F. (1983). Discussion. *Human Biology* **55**, 383–97.

Crow, J. F. (1989). Update to 'The estimation of inbreeding from isonymy.' *Human Biology* **61**, 949–54.

Crow, J. F. & Mange, A. P. (1965). Measurement of inbreeding from the frequency of marriages between persons of the same surname. *Eugenics Quarterly* **12**, 199–203.

Frisancho, A. R. (1979). *Human Adaptation: A Functional Perspective*. St. Louis: C. V. Mosby.

Howells, W. W. (1973). *Cranial Variation in Man: A Study by Multivariate Analysis of Patterns of Difference Among Recent Human Populations*. Cambridge, Massachusetts: Peabody Museum.

Howells, W. W. (1989). *Skull Shapes and the Map: Craniometric Analyses in the Dispersion of Modern Homo*. Cambridge, Massachusetts: Peabody Museum.

Jorde, L. B. (1982). The genetic structure of the Utah Mormons: Migration analysis. *Human Biology* **54**, 583–97.

Jorde, L. B. & Morgan, K. (1987). Genetic structure of the Utah Mormons: Isonymy analysis. *American Journal of Physical Anthropology* **72**, 403–12.

Lasker, G. W. (1977). A coefficient of relationship by isonymy: A method for estimating the genetic relationship between populations. *Human Biology* **49**, 489–93.

Lasker, G. W. (1985). *Surnames and Genetic Structure*. Cambridge University Press.

Lasker, G. W. & Kaplan, B. A. (1974). Anthropometric variables in the offspring of isonymous matings. *Human Biology* **46**, 713–17.

Lasker, G. W. & Kaplan, B. A. (1985). Surnames and genetic structure: Repetition of the same pairs of names of married couples, a measure of subdivision of the population. *Human Biology* **57**, 431–40.

Livingstone, F. B. (1958). Anthropological implications of sickle cell gene distribution in West Africa. *American Anthropologist* **60**, 533–62.

Livingstone, F. B. (1980). Natural selection and the origin and maintenance of standard genetic marker systems. *Yearbook of Physical Anthropology* **23**, 25–42.

Mascie-Taylor, C. G. N. & Lasker, G. W. (eds) (1988). *Biological Aspects of Human Migration*. Cambridge University Press.

Morton, N. E., Miki, C. & Yee, S. (1968). Bioassay of population structure under isolation by distance. *American Journal of Human Genetics* **20**, 411–9.

Pinto-Cisternas, J., Pineda, L. & Barrai, I. (1985). Estimation of inbreeding by isonymy in Iberoamerican populations: An extension of the method of Crow & Mange. *American Journal of Human Genetics* **37**, 373–85.

Relethford, J. H. (1988a). Effects of English admixture and geographic distance on anthropometric variation and genetic structure in 19th century Ireland. *American Journal of Physical Anthropology* **76**, 111–24.

Relethford, J. H. (1988b). Estimation of kinship and genetic distance from surnames. *Human Biology* **60**, 475–92.

Relethford, J. H. (1992). Analysis of marital structure in Massachusetts using repeating pairs of surnames. *Human Biology* **64**, 25–33.

Relethford, J. H. & Jaquish, C. E. (1988). Isonymy, inbreeding, and demographic variation in historical Massachusetts. *American Journal of Physical Anthropology* **77**, 243–52.

Roberts, D. F. (1978). *Climate and Human Variability*, (2nd edn). Menlo Park, California: Cummings.

Robins, A. H. (1991). *Biological Perspectives on Human Pigmentation*. Cambridge University Press.

Rogers, A. R. (1991). Doubts about isonymy. *Human Biology* **63**, 663–8.

Rogers, L. A. (1987). Concordance in isonymy and pedigree measures of inbreeding: The effects of sample composition. *Human Biology* **59**, 753–67.

Sorg, M. H. (1983). Isonymy and diabetes prevalence in the island population of Vinalhaven, Maine. *Human Biology* **55**, 305–11.

Swedlund, A. C., Meindl, R. S. & Gradie, M. I. (1980). Family reconstitution in the Connecticut Valley: Progress on record linkage and the mortality survey. In: *Genealogical Demography*, ed. B. Dyke & W. T. Morrill, pp. 139–56. New York: Academic Press.

9 A biological anthropological approach to measuring societal stress of parasitic disease: a case study of schistosomiasis

C. G. N. MASCIE-TALYOR & G. E. H. MOHAMED

Summary

Although Lasker's concept of plasticity is generally thought of in terms of growth, this chapter extends his approach to the study of the impact of infection with schistosomes. The chapter questions whether the World Health Organization is justified in suggesting that schistosomiasis is second only to malaria in socioeconomic and public health importance: schistosomiasis is not responsible for high mortality and its effects on work capacity and productivity, nutrition, and growth of children are not clear-cut.

Introduction

The extent to which humans show plasticity in response to stress (Lasker, 1969) is frequently discussed by biological anthropologists in relation to growth. This chapter focuses on the impact of disease, and in particular of one parasitic disease, schistosomiasis, and examines whether humans show plasticity in their response to disease.

Disease is generally thought of as an impairment of health and well-being whereby an individual falls below optimal functioning. In practice 'optimal' is not easy to define; good health status relies on a relatively subjective evaluation of the overall functional status of an individual within the limits set by the society within which s/he lives.

When discussing disease, three classes of contributory factors are usually recognized: the agent, the human host and the environment. The agent is the animate or inanimate proximal cause, essential for the disease to occur. Agents can be classified as nutritional elements, chemical agents, physiological, physical, genetic and psychic factors, and invading living organisms. Invading organisms belonging to the animal kingdom are generally called parasites; other organisms include bacteria and viruses. The host factors include both biological and behavioural ones which relate to susceptibility

159

or resistance. Biological factors include age, sex, ethnicity and specific immunity; behaviour includes habits and customs. The environment refers to all that is external to the agent and the human host; it encompasses the physical, biological and socioeconomic components of the environment. Causation of disease is typically complex and results from interaction between the agent, the human host and the environment.

There is increasing realization that there is a need to understand disease from a biological and anthropological perspective and to measure the resultant stress in society. The biological anthropological approach to the study of disease is holistic and incorporates an understanding of both the cause of the disease (epidemiology) and its impact.

The study of the impact of a disease incorporates Lasker's concept of plasticity. It is generally assumed that someone who is infected with, for instance, *Schistosoma mansoni* will suffer from ill-health. Is this assumption correct or can an infected person show plasticity in response and remain capable of leading an active and fully functioning life? This chapter attempts to address these questions by focusing on one parasitic disease, schistosomiasis, and determining its impact on fertility and fecundity, nutrition and growth, physical fitness, work capacity and productivity, genetic variation, epidemiology and cultural evolution.

Some authors and organizations (e.g. WHO, 1981) use the terms 'infection' and 'disease' interchangeably: that is, someone infected with *Schistosoma mansoni* would be said to be suffering from the disease schistosomiasis. Warren (1982) and others argue against this definition because they see an important distinction between infection and disease. In their view disease refers to a morbid state; infected people could be asymptomatic or diseased.

There are a number of reasons to support Warren's viewpoint, especially when discussing parasites. (1) The first definition implies that 'asymptomatic schistosomiasis' can exist; this is clearly not possible (it is a contradiction in terms). (2) Many parasitic infections do not give rise to specific symptoms and diagnosis is not possible. (3) In the case of schistosomiasis, disease results not from the number of copulating worm pairs but from the number of eggs produced by them.

Mortality and tropical diseases

International comparisons of mortality and morbidity are notoriously difficult. In many developing countries the true comparisons relate to the effectiveness or otherwise of the reporting system. Some governments are unwilling to publish such information, and in the past, data relating to plague, yellow fever, typhus, smallpox and famine have been suppressed.

Table 9.1. *Causes of mortality (all age groups) in developing countries*

Condition	Infections	Deaths
Respiratory disease	—	10 000 000
Circulatory disease	—	8 000 000
Diarrhoea	—	4 300 000
Measles	67 000 000	2 000 000
Injuries	—	2 000 000
Neoplasms	—	2 000 000
Malaria	2 600 000 000[a]	1 500 000
Tuberculosis	1 000 000 000	900 000
Hepatitis B	300 000 000	800 000
Whooping cough	55 000 000	600 000
Typhoid	70 000 000	600 000
Maternal mortality	—	500 000
Meningitis	—	350 000
Schistosomiasis	2200 000 000	250 000–500 000
Syphilis	15 000 000	200 000
Amoebiasis	500 000 000	70 000

[a]This figure refers to the population at risk inhabiting infected areas.

Nowadays governments are likely to suppress information about salmonella-type outbreaks and hepatitis, zoonotic events due to anthrax, and violence.

Various attempts have been made to produce tables of the causes of death in developing countries. Recently Walsh (1990) culled information primarily from World Health Organization publications. She found that the respiratory tract diseases, together with diarrhoeal diseases, cause between half and two-thirds of the mortality in developing countries (see Table 9.1). In other words, the major causes of death are frequently not 'the exotic parasitic diseases usually associated with tropical medicine, but are primarily bacterial and viral infectious diseases that were once endemic in industrialised countries but now are controlled through improvements in immunization, hygiene, nutrition, housing, water supply, and socio-economic status'. Thus schistosomiasis is not one of the major diseases causing mortality in tropical countries.

Life-cycle of schistosomiasis
Schistosomiasis is caused by schistosomes, which are digenetic dioecious trematodes. Humans may be infected by schistosome species of any of three groups: (a) major human parasites including *Schistosoma haematobium*, *S. mansoni*, and *S. japonicum*; (b) less epidemiologically prevalent species such

as *S. intercalatum* and *S. mekongi*; and (c) certain species of avian and mammalian schistosomes that produce dermatitis.

Schistosomiasis is an extremely widespread tropical disease, found in 76 countries (mainly Africa) where it affects over 200 million people with a further 400 million at risk. The disease has great longevity; it was symbolized in Egyptian hieroglyphics (3000 BC) by a dripping penis (indicative of blood loss in urine). Eggs of the urinary form of the disease, *S. haematobium*, have been found in the kidneys of twentieth-dynasty Egyptian mummies (1250–1000 BC). The parasite was first discovered by Theodor Bilharz in Cairo in 1851 after he noticed a schistosome worm during an autopsy. The name bilharziasis is in his honour.

The adult schistosomes are obligatory parasites; they exist in dioecious (separate) sexes in the definitive host (humans). The male worm is short (about 7 mm) and has a groove along the length of its body (the gynaecophoric canal) in which it carries a thinner but longer (up to 25 mm) female worm. Each worm has two suckers at its anterior end, with which it attaches itself to the blood vessels and sucks its blood meal. *S. haematobium* lives in the veins surrounding the urinary bladder and other pelvic organs; its eggs are voided in urine. *S. mansoni* inhabits the mesenteric veins of the colon and the eggs are excreted in faeces.

The life-histories of the schistosomes are similar for all species that cause patent infection in humans and involves an intermediate host (snail) and the final host (human). In humans, the adult female trematode produces eggs, which on reaching freshwater hatch into free-swimming miracidia (see Fig. 9.1). Each miracidium must find the proper snail host within 32 h or die. The miracidium is attracted chemotactically to the snail tissues. On finding the snail, it penetrates it and develops into a primary or mother sporocyst which later, through asexual divisions, produces between 200 and 400 secondary or daughter sporocysts. Each daughter sporocyst produces numerous fork-tailed cercariae, which escape from the snail and swim about in water.

A cercaria has up to 48 h in which to find a human host or die. On finding a host it penetrates the skin, during which process the tail is lost. The schistosomula, as it is then known, enters the blood circulation and is carried to the heart and then to the lungs. After some development in the lungs the trematode migrates to the liver to complete its growth. Fully grown schistosomes, with the male carrying the female in its groove, leave the liver and settle in the veins preferred by each species. There the females produce eggs and the life-cycle is repeated. The period from eggs to cercarial production takes between 4 and 7 weeks and the average period from cercarial penetration to egg excretion varies between 6 and 12 weeks.

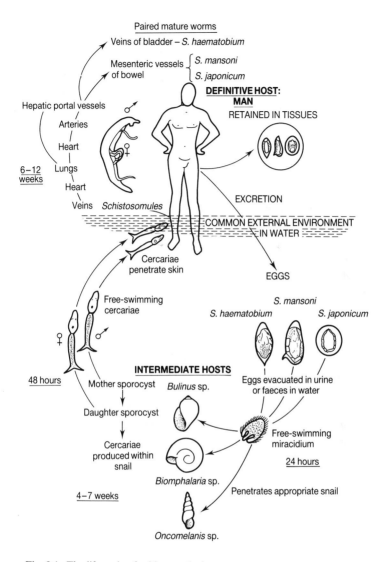

Fig. 9.1. The life cycle of schistosomiasis.

Schistosome worm pairs engage in egg-laying for many years; and their life-span has been estimated to average between 3.5 and 12 years with some worm pairs surviving for 30 years or more. Consequently schistosomiasis is a disease of long chronicity which ranges from first water contact during childhood well into adult life, albeit with decreasing prevalence after the second decade.

Schistosomiasis and morbidity

What are the consequences of infections with schistosomes? The major disease syndromes are related to the stage of infection. The first phase results from cercarial invasion; the dermatitis that results is thought to be due to host sensitization. Not all schistosomes appear to cause dermatitis (found with *S. haematobium* and *S. mansoni* but absent in *S. japonicum*). Dermatitis is almost always found in human infection with avian parasites and is probably a result of death of cercariae in the subcutaneous tissues.

Acute schistosomiasis or Katayama fever is a clinical syndrome that has been described a few weeks following primary infection, particularly with *S. mansoni* or *S. japonicum*. The clinical features include fever, hepatosplenomegaly, lymphadenopathy and peripheral eosinophilia. The aetiology of acute schistosomiasis is unknown but it is associated with heavy infection and has led to the notion that it is a form of serum sickness or antigen–antibody complex disease.

Most infected children have only minor early symptoms or none at all. They may continue in apparent good health during the subsequent chronic phase of schistosomiasis even while, in some individuals, lesions of the internal organs may be progressing. The lesions that ensue are caused by schistosome eggs rather than by worms. The adult copulating worms are impervious to the host immune system and thus are free of cellular reaction. Dead worms, which may appear in either the liver or lungs, give rise to focal lesions, but these are few in number and lack clinical significance compared with the millions of eggs generated in the course of infection.

Lesions resulting from chronic schistosomiasis can be divided into two major categories: local egg deposition and systemic cellular immune responses. Egg deposition occurs because the schistosomal eggs measure up to 70 μm in width and cannot cross capillary beds unaided. The eggs are laid in an immature state; it takes up to 10 days for schistosome eggs trapped in the host tissues to complete maturation of the enclosed miracidium, which can survive for a further 2–3 weeks. During this period several enzymes and antigenic materials are released, resulting in host sensitization. Cell-mediated hypersensitivity plays the key role in granuloma formation around *S. haematobium* and *S. mansoni* eggs, but the mechanism for the host's response to *S. japonicum* is still unclear.

Granuloma formation leads to a compact cellular infiltrate surrounding the schistosomal eggs; this is made up of lymphocytes, eosinophils, macrophages and fibroblasts. Formation of granulomas in the wall of the urinary or intestinal tracts is the major cause of pathological changes; granulomas can also lead to mechanical obstruction of the urinary tract or portal circulation.

In addition, granulomas can be transformed into permanent fibro-obstructive lesions although the switching mechanism(s) remain(s) undefined. In *S. mansoni* fibrosis occurs in the gut and the liver. In the liver, serious disease (pipe-stem fibrosis) can result from the restricted blood flow in the tissue surrounding the hepatic vessels. Clinical symptoms of severe urinary schistosomiasis in adolescents and adults can lead to blockage of urine flow, and sometimes fatal kidney disease may develop.

This brief review of the morbidity of schistosomiasis suggests that there is a sharp contrast between those individuals with mild schistosomiasis, which is of little consequence to the well-being and economic productivity of infected people, and the tragedy of advanced cases, which require hospitalization. Because the vast majority of infected individuals in endemic areas have been shown to have low intensities of infection, the health risk may well have been overemphasized.

Schistosomiasis and epidemiology

When epidemiologists discuss the impact of a disease they usually think in terms of the prevalence, intensity and incidence of the disease. In schistosomiasis, the prevalence would refer to the proportion of subjects who are infected at a given point in time. The intensity of infection indicates the worm burden but because it is so difficult to measure, the egg count (in either urine or faeces) is used as a proxy. There have been a few post-mortem cases where the worm burden has been counted. In a study in Brazil asymptomatic infected subjects had fewer than 10 *S. mansoni* worm pairs, but in 10% of the subjects there were between 160 and 320 worm pairs with one four-year-old having 1600 worm pairs (Cheever, 1968). The few studies that have attempted to show the relationship between worm burden and egg count indicate a complex association; a recent analysis indicates that the egg output from each worm is reduced in heavy *S. mansoni* infections (Medley & Anderson, 1985).

Prevalence and intensity of infection are related; in general, the higher the prevalence in a population the greater the mean egg count. However the relationship is by no means linear and plots of the prevalence against intensity of infection from a number of communities show that the best fit is a negative binomial distribution (see Fig. 9.2). There are two obvious conclusions to draw from the graphs: (1) most people in the community have low to moderate egg counts and only a few people have high burdens; (2) large changes in mean intensity are accompanied by only a small change in prevalence. Thus the assessment of control programmes (see later) should be based on changes in intensity of infection rather than prevalence.

The incidence rate measures the proportion of initially uninfected

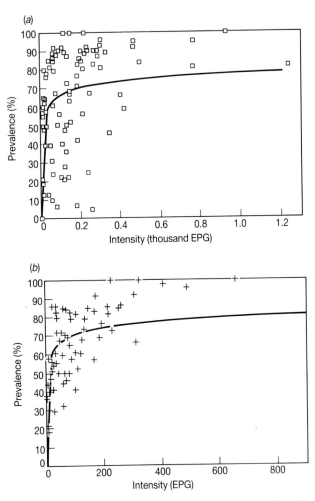

Fig. 9.2. The relationship between prevalence and intensity of infection for (a) *Schistosoma mansoni* and (b) *S. haematobium*. The solid line denotes the best fit relationship defined by the negative binomial ditribution.

subjects who become infected during a given period of time, usually expressed as the percentage per year. Incidence rates are difficult to measure (Scott, Senker & England, 1982); their main use is in monitoring control programmes where the reduction in incidence is a measure of their effectiveness.

There are very marked differences in the pattern of infection in human communities. The prevalence and intensity of infection for all the major forms of schistosomiasis varies with age and sex. In general there is an increase in prevalence and intensity from low levels in young children up to high levels in the 10–14 year age groups. Thereafter levels may fall slightly and stabilize before showing declines in 50–60-year-olds. This relationship has been defined as a convex form (Anderson, 1987). However not all populations conform to this pattern. Ongom & Bradley (1972) showed that the intensity of infection (for *S. mansoni*) in males increased with age in a fishing village in Uganda.

In many foci the prevalence of infection is greater in males than females, for example in Sudan, the Nile delta, the Philippines and northern Nigeria (Farooq *et al.*, 1966; Pesignan *et al.*, 1958; Pugh & Gilles, 1978). In the Gambia, Wilkins *et al.* (1986) found the prevalence and intensity to be very similar between the sexes despite very marked differences in exposure to infection. Occasionally the prevalence is higher in females, as in the Mende communities of Sierra Leone. In this population fishing is a female occupation; this explains the sex difference.

There are also discernible associations with occupation. A large study in the Nile delta showed that fishermen, boatmen, farmers and farm workers – all of whom were habitually exposed to water contact – had a high prevalence of infection (Farooq *et al.*, 1966). This study also demonstrated a significantly higher prevalence in Muslims than in Christians. Nordbeck, Ouma & Sloof (1982) showed that wealthier Kenyans in Machakos district had lower egg counts of *S. mansoni*, as did those with better farms. Finally, a study in Brazil (Kloetzel and Da Silva, 1967) has shown that the relationship between immigrants' length of residence in an endemic area and infection is similar to the relationship between age and infection in subjects born in the area. This finding suggests that length of exposure and infection, rather than a factor related to age, such as water contact, may be determining the pattern of infection.

An understanding of the pattern of water exposure is critical to epidemiology and successful control measures. Water contact can be classified under four main headings: occupational, domestic, recreational and religious. Adult males are usually the group involved in occupational exposure to infected water through activities such as fishing, farming and canal cleaning. In many endemic areas adult women spend a considerable time in domestic water-contact activities, including washing clothes and utensils and fetching water. Children frequently swim and play in water. Both children and adults regularly wash themselves in surface water, an activity which may be modified by the cercaricidal action of soaps. Contact with water for religious purposes is found in Muslim communities, where ablutions are carried out prior to praying.

Schistosomiasis and genetic variation

Unlike malaria, where at least five polymorphic genetic systems associate with the disease, very little is known of the relationship between schistosomiasis and genetic variation. Nevertheless, several genetic traits have been linked to the development of schistosomiasis. Individuals with blood group A are more likely to develop serious disease when infected with *S. mansoni*, whereas those of blood group O frequently develop mild disease (Camus *et al.*, 1977; Kamel *et al.*, 1978).

Two human studies have suggested a linkage of disease severity with specific HLA (human leukocyte antigen) haplotypes. These associations differed in Egypt (Abdel-Salam, Ishaak & Mahmoud, 1979) and in the Philippines (Sazazuki *et al.*, 1979). Researchers have also noticed that serious liver disease is more likely to develop in white than in black Brazilians infected with *S. mansoni* (Cheever *et al.*, 1978). Other have noted 'racial' variability in the degree and type of response to schistosomal infection (Jordan & Webbe, 1969; Gelfand, 1976).

Schistosomiasis and fecundity/fertility

The relationship between schistosomiasis and fecundity is controversial in both males and females. In the 1960s it was commonly reported that schistosomiasis caused 'mass chronic invalidism' but this view has now been discounted. As noted in an earlier section, even in areas where *S. mansoni* is hyperendemic, the majority of the population have light or moderate infections (as measured by egg counts) with no significant morbidity.

For schistosomiasis to have a significant effect on fecundity, schistosomal eggs must be found in the genital organs and it would be expected that the lesions formed would be sufficiently severe to affect adversely the functioning of these organs. Gelfand *et al.* (1970) in a male Zimbabwean sample of 100 autopsies found that 70 of the men had infected genital organs. In this study the eggs were mainly from *S. haematobium*. *S. mansoni* eggs are much less likely to be found in the genital organs. Even so, egg loads vary considerably; most studies have found that loads in the seminal vesicles are only about 20–25% of those found in the bladder. The testes and epididymis are rarely the sites of egg disposition.

More recent autopsy material indicates that male fecundity is rarely affected by schistosomiasis because the reactions to the eggs in the organs are very mild. McFalls & McFalls (1984) question whether mild lesions merely cause 'an imbalance in the intricate complex of biochemical and mechanical events necessary for full reproductive capacity'. They suggest that even though lesions might be mild they might play a more significant role during the early inflammatory stage. Furthermore, a heavily infected prostate gland could affect the quality and quantity of prostate fluid and the ejaculatory duct that passes through the prostate might be narrowed, giving rise to a reduced ejaculate volume.

Thus schistosomiasis might be a cause of temporary male subfecundity. In addition Abdul *et al.* (1974) (cited in Ledward, 1980) suggested that *S. haematobium* infections might cause male infertility because urogenital bilharziasis might enhance the autoimmune response, leading to a higher incidence of antispermal antibodies.

In females it has been claimed that chronic schistosomiasis causes multiple miscarriages and sterility (Retel-Laurentin, 1978). However, other researchers have argued that the eggs of *S. haematobium* are distributed throughout the female genital tract but substantial reactions to the eggs are confined to the lower tract (vagina and cervix) and therefore effects on fecundity are minimal.

The effects of schistosomiasis on the fallopian tubes has received considerable attention. However, in tubal schistosomiasis the mucosa of the tubes is rarely occluded (except in *S. mansoni* infections, which are rare). Instead the muscle layers are involved in the fibrotic process and this has led some researchers to propose that tubal schistosomiasis may be involved in ectopic pregnancy, as the thickened and damaged muscle layers cannot perform their function properly (Bland & Gelfand, 1970).

Von Lichtenberg (1987) reported that although schistosome eggs have been found close to the maternal–fetal interface and even in the amniotic fluid, there has been no documented instance of finding schistosome eggs in stillborn babies. Furthermore, no effects of schistosomiasis on birth weight or Agpar scores are known. However, tubal schistosomiasis may lead to pelvic inflammatory disease (PID), which can in turn result in ectopic pregnancy and tubal occlusion; this is one area where further work is required.

Schistosomiasis, nutrition and growth

There are three main ways in which parasites can cause maldigestion in humans: (a) by bringing about changes in the organs which secrete enzymes or in the ducts through which enzymes are secreted; (b) by impairing intestinal function through damaging the gut mucosa; (c) by reducing gut enzyme activity. Stephenson (1987) has provided a recent review of the relationship between schistosomiasis, malnutrition and growth. Infection with *S. haematobium* results in haematuria and proteinuria; anaemia and poor growth are widely reported in areas where urinary schistosomiasis is endemic. What is the urinary blood loss? Three studies of hospitalized patients (Gerritsen *et al.*, 1953; Farid *et al.*, 1968; Mahmood, 1966) with heavy chronic infections showed a daily loss ranging from 0.5 ml to 125 ml

Table 9.2. *Urinary iron loss: effects of treatment*

Group	Iron loss (μg per 24 h) pre-treatment			Subsample pre- vs. post-treatment					
				Pre			Post		paired
	n	mean	SE	n	mean	SE	mean	SE	t-test
Control	12	149	23	5	372	147	308	84	n.s.
Low–medium	19	278	66	11	341	72	226	77	$p = 0.086$
High	14	652	98	10	1250	269	344	54	$p = 0.002$
		$p < 0.001$			$p < 0.002$		n.s.		

per day; patients with severe or persistent haematuria lost 22 ml of blood on average per day. Losses of this magnitude, if persistent, are sufficient to increase dietary requirement of iron. Unfortunately none of these studies related severity of schistosomiasis to degree of blood loss.

In an attempt to determine the nutritional impact in non-hospitalized patients, Stephenson *et al.* (1985a–d, 1986) studied urinary iron loss in a sample of 45 Kenyan primary school children. The children were divided into three groups, low to medium (16–177 eggs per 10 ml), high (200–1194) and uninfected (control). Before treatment the control group was losing, on average, 149 μg of iron per 24 hours, the low to medium group 178 μg, and the high groups 652 μg. As can be seen from Table 9.2, iron loss was significantly greater in the high-egg-count group than in the control and low to medium groups. There was a significant linear relationship between iron loss per 24 hours in the infected children but the coefficient of determination (R^2) was 0.32, which indicated a considerable amount of individual variation in iron loss not due to the intensity of infection.

The infected children were treated. Seven weeks later iron losses in the previously low–medium and high groups were similar to those of the control group; reductions in iron loss were concomitant with reduced egg counts, down by close to 99% in both infected groups (Table 9.2). Iron loss was therefore associated with reversible bladder lesions, which healed within a seven-week period.

The relationship between *S. haematobium* infection and anaemia has generally been reported from cross-sectional studies and thus conclusive longitudinal evidence is still awaited. For instance, Abdel-Salam & Abdel-Fattah (1977) compared the haemoglobin levels of 800 Egyptian children and found that heavily infected children (those with over 350 eggs per 10 ml) had a significantly lower mean value (8.5 g dl^{-1}) compared with uninfected children (9.8 g dl^{-1}). Children with moderate infections, defined as between 100–350 eggs per 10 ml, had an intermediate mean value of 9.1 g dl^{-1}, but this was not different from the other two means.

However, four cross-sectional studies (two in Zambia, and one each in Zimbabwe and Liberia) all failed to find any relationship between infection and anaemia, although in all four studies the sample sizes were small and no measurement of intensity of infection was obtained (Golden & Barclay, 1972; Bell *et al.*, 1973; Friis-Hansen & McCullough, 1961; Holzer *et al.*, 1983). A Nigerian cross-sectional study in school children also found no relationship between anaemia and intensity of *S. haematobium* infection but less than 5% of the children were suffering from moderate or severe anaemia (defined by haemoglobin less than 10 g dl^{-1}).

Stephenson *et al.* (1985c) undertook a multiple parasitic (*S. haematobium*, hookworm and malaria) infection study on Kenyan children. Using multiple regression analysis, they were able to show that the decrease in *S. haematobium* and hookworm egg counts were of equal importance in haemoglobin rise after treatment. These treatment changes in the case of *S. haematobium* are thought to be due to a decrease in abnormal iron loss in urine. Experiments on mice (Mahmoud, 1980, 1982) also suggest that iron loss results from increased red cell destruction by the spleen and decreased cell production. Support for these findings come from Stephenson's results (1985a), which showed that the increase in haemoglobin levels after treatment was associated with a decrease in splenomegaly and mean spleen size.

A number of studies have examined the relationship between protein loss and *S. haematobium* infection. For instance, Wilkins *et al.* (1979) in a Gambian study found that 54% of those with egg counts of 100–999 eggs per 10 ml urine had proteinuria of 100 mg per 100 ml urine or more and 12% of the sample had proteinuria of 300 mg per 100 ml. If 1 l of urine is voided per day then daily urinary protein loss is 1 g in 54% of the sample and about 3 g in 12%. However, Norden & Gelfand (1973) concluded from a Zimbabwean study that protein loss from *S. haematobium* would be important only in extreme cases of kwashiorkor and marasmus: they found only 1 case (out of 130 children) where protein excretion exceeded 1 g per 24 h. The mean amino acid excretion was 47 ± 45 mg of α-amino nitrogen per day, which was slightly below the normal values.

Finally a large number of studies have investigated the relationship between *S. haematobium* infection, child growth and adult body mass index (weight in kilograms divided by (height in metres) squared). Four cross-sectional studies in Africa have found no relationship between infection and children's anthropometric measurements. However, these studies have limitations: three of them did not measure intensity of infection, two of them involved small sample sizes, and in one study the prevalence of infection was low.

On the other hand six African studies have suggested that urinary schistosomiasis may inhibit child growth and decrease adult weight for height. In an Egyptian study of 800 children (Abdel-Salam & Abdel-Fattah, 1977) the heavily infected (>350 eggs per 10 ml) had significantly lower skinfold thicknesses for age compared with uninfected children: weight, height and arm circumference were also lower than in the uninfected children, but not significantly so. In Northern Nigerian adolescents and adults, Oomen, Meuwissen & Gemert (1979) found that *S. haematobium* was significantly associated with lower body mass index.

These conflicting results suggested that there was a need for longitudinal rather than cross-sectional studies with sufficiently large sample sizes to control for confounding effects due to differences in age, sex, diet, socioeconomic status and polyparasitism. Stephenson *et al.* (1985b, 1986) undertook a longitudinal study of Kenyan primary school children in an area where *S. haematobium* and hookworm were common, malaria was endemic and where there was poor growth. Initially (time 1) all children with light to moderate *S. haematobium* infections (defined as 1–500 eggs per 10 ml) were randomly assigned to either a placebo ($n = 198$) or treatment group ($n = 201$). Both groups were treated six months later and re-examined. The treated group showed more rapid growth between time 1 and time 2 than did the placebo group. Indeed the placebo group showed significant declines in percentage weight for age, percentage arm circumference for age and skinfold thicknesses, and no significant increase in percentage height for age. On the other hand the treated group showed increases in all anthropometric measurements and for nearly all these measurements treated children grew better (see Table 9.3). Growth benefits due to treatment were maintained for up to 16 months after the first metrifonate treatment.

The mechanisms by which treatment for *S. haematobium* infection could improve growth rates in children is still unclear. Although *S. haematobium* causes proteinuria the amount of protein lost per day is unlikely to be of major importance. Some researchers (Cheever, 1985; Stephenson, 1987) have speculated that treatment triggers an improvement in appetite and thus an increase in food intake. Measuring food intakes is notoriously difficult in humans but support for the hypothesis comes from non-human primates, where anorexia has been found in acute *S. haematobium* infections.

There have been far fewer studies of the association between *S. mansoni* and malnutrition. Although it is generally acknowledged that *S. mansoni* causes blood loss in the stools, the magnitude of the loss and its effects on

Table 9.3. *Relationship between selected anthropometric measurements and* S. haematobium *infection in placebo and treated groups*

Measurement	Group	Time 1		Time 2		paired t-test	t-test (placebo vs. treated)
		mean	SE	mean	SE		
Weight	Placebo	30.0	0.51	31.6	0.56	0.001	0.001
(kg)	Treated	29.1	0.52	31.5	0.56	0.001	
Height	Placebo	138.2	0.82	141.0	0.83	0.001	0.001
(cm)	Treated	137.1	0.82	140.1	0.82	0.001	
Arm	Placebo	18.2	0.15	18.3	0.16	0.001	0.001
circumference	Treated	18.0	0.16	18.6	0.17	0.001	
(cm)							
Triceps	Placebo	7.7	0.19	7.7	0.20	n.s.	0.001
skinfold	Treated	7.5	0.17	8.4	0.18	0.001	
(mm)							
Subscapular	Placebo	6.0	0.12	6.1	0.14	n.s.	0.001
skinfold	Treated	6.0	0.13	6.8	0.14	0.001	
(mm)							

nutritional status remain unclear. Studies of faecal blood loss on small numbers of Egyptian patients with colonic and rectal polyps (caused by *S. mansoni*) ranged from 7.5 to 29 ml per day (mean 12.5 ml), equivalent to between 0.6 and 6.7 mg (mean 3.3 mg) of iron per day (Farid *et al.*, 1967; Farid, Patwardhan & Darby, 1969). In a later study of blood loss in 15 Egyptian farmers with colonic polyposis the mean was 13.0 ml, equivalent to a mean daily iron loss of 4 mg (Waslein, Farid & Darby, 1973). Other studies of patients with chronic heavy infections with or without colonic polyps have reported abnormally high losses of albumin, zinc and vitamin A as well as decreased D-xylose absorption, elevated fat excretion and impaired glucose tolerance (El Rooby *et al.*, 1963).

Eleven studies have examined whether *S. mansoni* infection is associated with anaemic status (based on haemoglobin concentration), growth, and physical fitness, as well as other measures of nutritional status such as serum albumin levels. Salih, Marshall & Radalowicz (1979) studied 511 Sudanese males infected with *S. mansoni* and found the heavily infected ones (defined as > 400 eggs/g faeces) to have significantly lower mean haemoglobin concentrations (by 0.9 g dl^{-1}) than more lightly infected individuals. In an Egyptian study, Mansour, Francis & Farid (1985) measured iron status and found 50% of the sample ($n = 103$) to be anaemic with haemoglobin concentrations below 13 g dl^{-1}. In Tanzanian school children those infected with *S. mansoni* had significantly lower haemoglobin concentrations as well as lower weights and heights than their uninfected colleagues (Jordan & Randall, 1962).

Other studies have shown no significant difference between infected and uninfected individuals. For example a Ugandan (Ongom & Bradley, 1972) study found that infected villagers did not show significantly lower haemoglobin concentrations than uninfected ones (although infected villagers with ascites did show much lower mean haemoglobin than the community as a whole, 8.7 g dl^{-1} versus 10.7 g dl^{-1}). Studies in Puerto Rico (Cline *et al.*, 1977) and Brazil (Lehman *et al.*, 1976) failed to show significant differences in haemoglobin concentrations between infected and uninfected individuals; a St Lucia study (Cook *et al.*, 1974) found no difference in anthropometric measurements even when intensity of infection was taken into account. In the latter study heavily infected children showed significantly lower serum albumin levels than matched (by sex and age) uninfected children.

A recent Kenyan study (Corbett *et al.*, 1992) has suggested a possible association between nutritional status and *S. mansoni* infection although the relationship between intensity of infection and anthropometric measurements was complex and not easily interpretable. The authors hypothesize that schistosome associated morbidity leads to a subsequent nutritional defect which can be tested by appropriate intervention studies.

There have been very few studies of the relationship between *S. japonicum* and nutritional status (or so we assume; it could be that many have been published in China and that we are simply unaware of them). Those that have been published involve either small clinical studies or morbidity studies where no direct nutritional measurements were obtained.

In conclusion the relationship between schistosomiasis and nutritional status remains unclear. This is because the disease is found in those parts of the world where other parasitic infections, malnutrition, poverty and poor hygienic conditions are prevalent and the relative importance of schistosomiasis *per se* is difficult to evaluate. In addition, schistosomiasis is not characterized by acute, readily recognized episodes of disability which can be counted and measured.

Schistosomiasis, physical fitness, work capacity and productivity

It is generally assumed that infected children and adults are debilitated, with a consequent reduction in ability to work. In one of the first studies Okpala (1961) in Nigeria emphasized the lethargy of school children infected with *S. haematobium*. He stated: 'the disease affected the physical well-being of the pupils, reducing their strength for work and retarding their growth and mental development. Further observations revealed that the majority of the school pupils could hardly finish 3.5 minutes of physical exercise (running or jumping) without showing signs of extreme fatigue and physical exhaustion'. These findings imply that there were very obvious differences in the physical performance between infected and non-infected children; this means that there should be considerable improvement in physical activity when infected individuals are treated.

However, results from the 1960s and 1970s failed to support Okpala's views. For example, Jordan & Randall's (1962) study of Tanzanian school children using a step test to measure physical fitness found no significant differences between infected and uninfected children, although the sample sizes were rather small and no attempt was made to determine intensity of infection. Walker *et al* (1972) failed to find any significant differences in physical performance of Bantu children using a 12 minute walk and run test; and Davies (1972) also obtained negative results using a bicycle ergometer. Both of these studies have been criticized by Stephenson (1987) on the grounds that intensity of infection was not taken into account, the sample size of the infected group was only ten (Davies 1972) and the walk–run test does not separate work load and motivation.

Stephenson *et al.* (1985c,d) conducted a similar study to Jordan & Randall using Kenyan primary school children but did measure intensity of *S. haematobium* infection and matched them with uninfected (control) children for potential confounding variables. They found that infected children were less physically fit than controls and had higher heart rates after completion of the Harvard Step Test than did the uninfected controls. After treatment the previously infected children showed improved performance and their levels of physical activity were similar to those of the uninfected group. This result suggests that the diminished physical performance of infected children is easily reversible by treatment.

How schistosomiasis affects physical fitness is still a matter of conjecture. As noted earlier *S. haematobium* can give rise to iron deficiency, anaemia and protein–energy malnutrition as well as general malaise. Because Stephenson *et al.* matched their samples for haemoglobin concentration and anthropometry, the initial differences in step test performance cannot be due to these factors.

Of related interest is whether schistosomiasis in children is associated with school performance, mental ability or cognitive skills, and school absenteeism. Relatively few studies have been undertaken. Abdalla, Badran & Galal (1964) used psychometric tests to study the performance of 302 Egyptian school children. Although the infected children scored lower, on average, than uninfected children the differences were not significant, with the exception of the numerical ability subtest. Walker, Walker & Richardson (1970) used examination results as a proxy for academic performance but found no significant difference in results between infected and uninfected children. Weisbrod *et al.* (1974) also found no significant difference between infected and uninfected children when ranking the child's position in the class.

Absenteeism from school has also been studied. Walker *et al.* (1970) found no evidence for any relationship between school attendance and infection. On the other hand, Weisbrod *et al.* (1973, 1974) in a St Lucia study found a significant relationship (at the $p \leq 0.05$ level) but infected children were less likely to be absent from school than uninfected school children. This paradoxical result is probably because healthy children were kept from school to work in the fields. Thus an infected child might be too weak to work in the fields but well enough to attend school.

Absenteeism has also been studied in adults as a means of assessing the economic consequences of infection. The results, which mainly pertain to *S. mansoni* infection, have been reviewed by Prescott (1979). Many of the studies were conducted between the late 1940s and the early 1970s. Khalil (1949) felt that up to £80 000 a year was lost in Egypt due to schistosomiasis; Wright (1972) estimated that the productivity lost in Egypt through schistosomiasis was 33%.

These figures have been criticised by Prescott (1979), Weisbrod *et al.* (1973) and Gwatkin (1984) because they assume that (a) clinical symptoms can be directly equated with lost production (b) all infected workers will show the same loss in productivity and (c) after treatment the workers will increase their output by an equal amount. The calculations do not take into account the size of the labour force or the capital outlay required in order to increase productivity.

More recent field studies have attempted to overcome some of these difficulties. Thus Foster (1967) examined the health and productivity records over one year (1962–63) of infected and uninfected workers on a sugar estate in Tanzania. He found no significant difference in output between infected and uninfected workers but did show that infected workers had higher levels of absenteeism and were more likely to require medical treatment. A follow-up study five years later in the same sugar plantation by Fenwick & Figenschou (1972) showed retrospectively that the mean bonus earnings of uninfected cane cutters exceeded those of infected workers in four consecutive six-month periods (between January 1968 and December 1969) by 11%, 11.4%, 6.0% and 13.4%.

Fenwick & Figenschou (1972) also made a time-series study in which they compared the bonus earnings of an infected group of cane cutters before and after treatment with those of untreated cane cutters as well as with those of an uninfected group. They found that the mean bonus earnings of the treated group improved in comparison to both the untreated and uninfected groups.

Physical fitness, absenteeism and productivity are not the only measures used to study the impact of an endemic disease on the ability to work. Physiologists usually measure energy expenditure. There are two important components to energy expenditure: (a) potential (or maximal) work, which can be undertaken under controlled (i.e. laboratory) conditions; and (b) the actual work performance in a field situation. Both of these measures beg the question of whether either or both reflect accurately the habitual energy expenditure which would be expected under normal occupational activities.

A number of methods exist to measure the physiological component of physical work. One method is to determine maximum aerobic capacity (V_{O_2max}) using ergonomic tests. Unfortunately this method necessitates an artificial work situation and is laboratory-based. Even so, 'V_{O_2max} is regarded as a standard index of overall fitness or work capacity' (Collins *et al.*, 1988). There are methodological limitations; in particular, the determination of maximal aerobic power is usually obtained only from trained athletes. For the less fit subjects submaximal intakes are used to predict maximal levels although such results can be confounded by age and habituation effects.

Van Ee & Polderman (1984) studied the impact of *S. mansoni* on work capacity of 160 tin miners from four villages in the district of Maniema, Zaire. Each miner undertook submaximal work tests using both a bicycle ergometer and step test. No significant differences in mean work capacity were found between infected and uninfected miners and there was no indication of reduced physical fitness. However, the intensity of infection was low and in the two most heavily infected villages only 29% and 27% of the miners had faecal egg counts of 600 eggs g^{-1}. The geometric mean count was also low; 210.5 and 251.1 eggs g^{-1}, respectively.

Collins and his colleagues have undertaken a number of studies of energy expenditure in Sudan (Awad El Karim *et al.*, 1980; Collins *et al.*, 1976). Their initial study was on the work performance of cane cutters living in two villages in a sugar plantation on the Blue Nile south of Khartoum. After an initial screening of 500 men aged between 16 and 60 years, those infected with *S. haematobium*, malaria, other parasitic infections or with haemoglobin concentrations below 10 g per 100 ml were excluded. In total 194 male cane cutters were studied, divided into younger (16–24 years) and older (25–45 years) age groups. Each age group was further divided into three groups: (1) uninfected (no *S. mansoni* eggs excreted in faeces): (2) infected (irrespective

Table 9.4. *Mean oxygen uptake in three groups of Sudanese cane cutters*

Not infected		Submaximal V_{O_2}			V_{O_2}			
					absolute		relative weight	
		(1 min^{-1})			(1 min^{-1})		$(\text{ml kg}^{-1} \text{ min}^{-1})$	
	n	mean	SE		mean	SE	mean	SE
Younger age group								
Not infected	28	2.23	0.16		2.88	0.43	48.3	5.7
Infected	40	2.16	0.22		2.84	0.57	50.7	7.1
Infected + clinical signs	31	2.26	0.19		2.83	0.32	49.8	5.4
Older age group								
Not infected	27	2.13	0.19		2.67	0.42	47.3	6.4
Infected	41	2.20	0.14		2.80	0.49	46.4	7.6
Infected + clinical signs	27	2.21	0.16		2.97	0.70	49.3	8.7

of egg count) and without clinical signs of hepatomegaly; and (3) infected with clinical signs.

Each subject performed a progressive exercise test on a bicycle ergometer. During the final minute of each work load the minute ventilation, oxygen intake and cardiac and respiratory frequencies were measured. Measurement of $V_{O_2\text{max}}$ was not attempted but was estimated from the regression of oxygen consumption and heart rate. The results (Table 9.4) showed no significant group differences for either submaximal oxygen uptake or predicted maximal oxygen uptake (absolute or relative to body weight). Thus the laboratory data provided no evidence that *S. mansoni* infection impairs physiological response to a standard test of physical work.

Collins *et al.* (1976) also measured individual productivity (weight of cane per unit time) in 129 of these cane cutters with representative samples drawn from each of the six groups. Details of the field study techniques have been provided (Collins *et al.*, 1976) but the aim was to measure productivity both on normal days and when a monetary bonus system was used as an inducement. The results were not in accord with expectations; in general the non-infected were the lowest producers, the infected group the highest and the infected group with clinical signs intermediate. There was a 10% difference in production between infected and non-infected and a 5% difference between infected with clinical signs and non-infected. The bonus scheme increased productivity in all three groups but the infected vs. uninfected difference was maintained.

Collins *et al.* (1976) suggested that these paradoxical results were due to differences in work experience of the three groups since they had shown that overall there was a highly significant positive relationship between

Table 9.5. *Mean differences in aerobic response and haemoglobin concentrations between non-infected and infected villagers and canal cleaners in Sudan*

		V_{O_2}						
		absolute ($1\ min^{-1}$)		relative weight (ml kg^{-1} min^{-1})		Hb (g dl^{-1}		Egg excretion
Group	n	mean	SE	mean	SE	mean	SE	mean	SE
Non-infected villagers	37	2.71	0.71	46.9	12.0	15.2	1.3	0	—
Infected villagers	146	2.63	0.56	47.4	10.7	14.9	1.2	367	486
Canal cleaners	19	2.39	0.46	41.9	5.7	13.3	2.2	2054	1105

experience and mean productivity. The uninfected group had on average 1.4 ($\pm$ 1.5) seasons' experience, whereas the infected without and infected with clinical signs groups had 3.5 ($\pm$ 2.4) and 3.6 ($\pm$ 2.5) seasons' experience, respectively.

In a second study (Awad El Karim *et al.*, 1980) a settled village community was used to measure the physiological responses to physical exercise using the same procedures as before. Only the economically active males in the age group 18–45 years were studied and an anti-malarial prophylaxis programme was arranged to ensure that none of the subjects tested was suffering from malaria. In total 203 males participated of which 37 were uninfected villagers, 147 were infected villagers and the remainder (19) were cleaners of irrigation canals. There was a highly significant difference in egg counts between infected villagers (mean 367 $\pm$ 486) and the canal cleaners (mean 2054 $\pm$ 1105).

The results showed that there were no significant differences between the two village groups, i.e. low *S. mansoni* infection did not impair physical performance. However, there was a highly significant difference in maximum aerobic output (V_{O_2max}) between villagers and canal cleaners (Table 9.5). This impairment of up to 18% was apparent when the intensity of infection increased from light (defined by the authors as fewer than 1000 eggs g^{-1}) to very heavy (more than 2000 eggs g^{-1}). Again these results must be treated with caution since (a) the canal cleaners are immersed up to the waist in water and it is possible that this might cause 'detraining of the leg muscles for work tests on a bicycle ergometer' (Collins *et al.*, 1988); (b) the canal cleaners had a significantly lower mean haemoglobin concentration than villagers and this might partly account for the reduced work capacity (Table 9.5).

In a more recent study in Sudan, Awad El Karim *et al.* (1981) assessed the effect of treatment on physical work capacity. Twenty-two males infected with *S. mansoni* were selected but none of the subjects had splenomegaly or hepatomegaly before or during the experimental period. After the initial laboratory investigations in which the same pulmonary function, anthropometric and haemoglobin tests as described earlier were used, the subjects were treated with hycanthone (Etrenol). After release from hospital the subjects were instructed to avoid reinfection. Stool samples were taken every three months to ensure that the subjects had remained uninfected. After a period of between nine months and one year the subjects were reassessed. In addition an untreated control group of 19 infected villagers were also examined on two occasions, four and six months later.

There were no significant differences in changes in body weight, lean body mass and leg volume between treated and control groups, but the treated group increased significantly in mean body weight ($+3.3$ kg) and lean body mass ($+2.4$ kg) between the two tests. Mean egg counts fell from 526 ± 428 eggs g^{-1} to 0 in the treated group and increased from 257 ± 268 eggs g^{-1} to 445 ± 395 eggs g^{-1} in the untreated controls. Haemoglobin concentrations also changed significantly between the two groups, with an increase of 1.1 g per 100 ml in the treated group compared with a smaller increase of 0.5 g per 100 ml in the controls.

The pulmonary function tests showed that there was only a small change in submaximal aerobic response after treatment, but $V_{O_2 \text{max}}$ increased significantly after treatment, as did forced vital capacity. In addition half of the treated subjects reported that they suffered far less fatigue in their everyday work.

Finally Parker (1989) has examined the effects of *S. mansoni* infection on female activity patterns in the Gezira–Managil area of Sudan. She undertook two separate matched trials. In the first she measured the daily activity patterns and productivity (cotton picking) of 22 females; the eleven infected women had a mean egg load of 676 eggs g^{-1} and the matched group were uninfected. Parker took particular care with matching, which took into account age, socioeconomic status, household composition (including domestic labour) and where the cotton was picked (whether on her own rented tenancy, a family member's tenancy or one where she was employed which had no strong familial links). With this precise matching she hoped to overcome the deficiencies of the small sample sizes.

Minute-by-minute observations were carried out both in the morning and afternoon cotton picking sessions. The pairs were, wherever possible, observed on consecutive days so as to minimize variation due to seasonal factors. Parker found that there was no difference in absenteeism in the morning session between the two groups but infected women were less likely to return to the cotton fields in the afternoon; infected women also spent less time engaged in agricultural physical work activities in the morning and afternoon. However, no significant differences were found in amount of cotton plucked if the absent women were excluded from the analysis. If all pairs were included then there was a significant decrease in productivity in the infected women.

The second matched trial was of 12 pairs, and observations were restricted to domestic work activities. Here she found that there were no differences in the time engaged in domestic work (light versus heavy) between infected (mean egg count 574 eggs g^{-1}) and non-infected women, nor did infective status affect the time spent engaged in infant and child care activities. Why *S. mansoni* infection should have a greater impact on cotton picking and less in the domestic sphere remains largely unanswered, although Parker suggests that cotton picking involves direct exposure to the sun and thus cotton picking creates two loads, work and an additional heat load.

This section also shows that the impact of infection on physical fitness and energy expenditure is not straightforward. Many of the conflicting results can be explained by differences in study design and methodology and the inability to take into account confounding variables such as experience. In addition only reasonably fit and healthy people participate, especially in the more physiologically orientated tests.

Schistosomiasis, behaviour and the implications for control programmes

There is no doubt that human behaviour is partly responsible for the continuing spread of the disease. Europe first learned about schistosomiasis after Bonaparte's expedition into Egypt in 1798 when his soldiers soon began to suffer severely from a disease characterized by the painful passing of bloody urine. A century later schistosomiasis became such a problem to British troops serving against the Boers that for years the British government annually paid out £10 000 to soldiers incapacitated by the disease. When the Americans stormed ashore on Leyte in October 1944 they encountered these same parasites, about which they had received no warning. It was not until New Year's Day 1945 that the first military cases were diagnosed among patients in an evacuation hospital. In the end 1700 men were put out of action by the fluke at a cost of 300 000 fighting man days and three million US dollars. The lesson of the dangers from bilharzia had not been learned five years later when 50 000 Chinese communist soldiers assembled during 1950 for the projected invasion of Formosa. Schistosomiasis became widespread among them and led to the abandonment of the campaign and the survival of Chiang Kai-Shek's island nation of Taiwan.

Schistosomiasis is spreading because of the dramatic increase in snail habitats created by water resource developments, both large man-made lakes for irrigation schemes and generating hydroelectric power, and small dams to conserve water for use by humans and their livestock. Such developments inevitably attract increasing human populations, which enhance the spread and increased transmission of schistosomiasis.

For many years the World Health Organization (WHO) took the view that the snail was the culprit. Then in 1984 WHO announced that schistosomiasis is caused by people, not snails. The acceptance of this idea has caused and is causing a revolution in the concept of schistosomiasis control.

Control involves interruption or reduction in the transmission of the parasite either by halting the introduction (or reintroduction) of infection into the community or by prevention of exposure to cercarial infected water. Therefore the aims of control programmes focus on trying to reduce infection in water by the use of focal mollusciciding to kill infected snails; reducing contamination by chemotherapy and provision of latrines; reducing exposure by improvement in water supplies; reduction in water contact by building fences and bridges; and finally by educating people about the disease and how it can be prevented. Even so there has been concern over the recurring introduction of chemicals into fresh water and the rising costs of chemotherapy.

A number of large-scale control programmes have been executed throughout the world. For instance, in Brazil a special programme for schistosomiasis started in 1976 in eight northeastern states where nearly five million people were at risk. The programme treated school-aged children and used focal mollusciciding, health education and provision of clean water supplies and sanitation facilities. As a result the programme achieved its objective of reducing the intensity of infection to single figures. Other programmes in Egypt and the Philippines have had only limited success.

The project of which the authors have personal knowledge is the Blue Nile Health Project of Sudan. Sudan is the largest country in Africa but much of it is desert or semiarid flat plains. The country's major asset is the Nile; since 1924, with the completion of the Sennar Dam, improved water management has led to increasing use of fertile Nile flood plains. The famous Gezira scheme was opened in 1925 with 250 000 acres of land irrigated via a network of open canals. Today the Gezira–Managil scheme totals two million acres and the cotton it produces generates over 50% of total export earnings. But the price paid for agricultural development was a dramatic increase in both the prevalence and intensity of schistosomiasis: over half of the residents of these agricultural areas were thought to be infected, and in school children the prevalence was over 80%.

In the early 1980s new anti-schistosomal drugs were being developed; praziquantel became the first drug considered safe to use on a large scale. In Sudan, praziquantel is used for mass treatment in villages with prevalence rates of more than 40%; in villages where the prevalence rate was lower only infected persons were treated (selective population chemotherapy). Further measures taken include focal mollusciciding, weed removal in some areas, building of pit latrines, provision of safe water supplies and the introduction of health education programmes.

At the start of the programme in 1984 the prevalence of *S. mansoni* varied in different areas of the Gezira–Managil. This variation could be explained by varying levels of hygiene, education and type and condition of the water supply. In those villages which had good piped water supply and ground water, the prevalence was about 50% whereas villages with little or no health and sanitation facilities, and which relied on canals for their water supply, had prevalences close to 90%.

The control measures were undertaken in phased annual programmes and the progress was monitored by annual surveys of 5000 school children. Mohamed (1992) used logistic regression analysis to examine the effectiveness of the different components of the programme on changes in prevalence. He was able to show that the logistic regression models correctly predicted between 71% and 90% of positive infections and between 10% and 60% of negatives.

Multiple regression analysis was used to monitor changes in the intensity of infection. The main variable responsible for reduction in egg count was chemotherapy (treatment). However, the geometric mean was reduced further if a treated child was using tap water, and further still if there was a latrine in the home. The coefficient of determination (R^2) ranged from 10% to 23% over the five annual surveys. Restricting analyses to positives only resulted in poorer prediction with the coefficient of determination ranging from 5% to 10%. The results of these analyses suggest that the comprehensive strategy adopted in Sudan is working, leading to a marked reduction in prevalence and intensity of infection in school children. There is a need for continued external and governmental assistance to ensure that reductions in prevalence and intensity of infection are sustained.

Conclusions

Parasitic diseases are generally thought of as a 'stress'. This review has shown in relation to schistosomiasis that the consequences are quite difficult to assess. Schistosomiasis does not seem to greatly affect fecundity and fertility; it is not responsible for high mortality; and its effects on nutrition, growth, work capacity, physical and mental fitness, and productivity are still open to doubt. It is difficult to justify the comments made by a panel of experts from the World Health Organization (1985) who wrote: 'In terms of socioeconomic and public health importance in tropical and subtropical areas, it (schistosomiasis) is second only to malaria'.

What we can say with certainty is that schistosomiasis is largely caused by human behaviour: in this case, principally water use practices and indiscriminate urination and defecation, but also failure to take advantage of available screening practices or to comply with medical treatment. However, it is important to avoid blaming people and their behaviour exclusively, especially when appropriate control measures are not widely available.

References

Abdalla, A., Badran, A. & Galal, S. (1964). The effect of bilharziasis on the mental power and scholastic achievement of school children. *Journal of the Egyptian Public Health Association* **39**, 135–46.

Abdel-Salam, E. & Abdel-Fattah, M. (1977). Prevalence and morbidity of *Schistosoma haematobium* in Egyptian children. *American Journal of Tropical Medicine and Hygiene* **26**, 463–9.

Abdel-Salam, E., Ishaak, S. & Mahmoud, A. A. F. (1979). Histocompatibility linked susceptibility for hepatosplenomegaly in human schistosomiasis mansoni. *Journal of Immunology* **123**, 1829–31.

Anderson, R. M. (1987). Determinants of infection in human schistosomiasis. In: *Baillière's Clinical Tropical Medicine and Communicable Disease*, 2nd ed., ed. A. F. Mahmoud, pp. 279–300. London: Baillière Tindall.

Awad El Karim, M. A., Collins, K. J., Brotherhood, J. R., Dore, C., Weiner, J. S., Sukkar, M. Y., Omer, A. H. S. & Amin, M. A. (1980). Quantitative egg excretion and work capacity in a Gezira population infected with *Schistosoma mansoni*. *American Journal of Tropical Medicine and Hygiene* **29**, 54–61.

Awad El Karim, M. A., Collins, K. J., Sukkar, M. Y., Omer, A. H. S., Amin, M. A. & Dore, C. (1981). An assessment of anti-schistosome treatment on physical work capacity. *Journal of Tropical Medicine and Hygiene* **85**, 65–70.

Bell, R. M. S., Daly, J., Kanengoni, E. & Jones, J. J. (1973). The effects of endemic schistosomiasis and of hycanthone on the mental ability of African school children. *Transactions of the Royal Society of Tropical Medicine and Hygiene* **67**, 694–701.

Bland, K. & Gelfand, M. (1970). The effect of schistosomiasis in the fallopian tubes in the African female. *Journal of Obstetrics and Gynaecology of the British Commonwealth* **77**, 1024.

Camus, D., Bina, J., Carlier, Y. & Santoro, F. (1977). A, B, O blood groups and clinical forms of schistosomiasis mansoni. *Transactions of the Royal Society of Tropical Medicine and Hygiene* **71**, 182.

Cheever, A. W. (1968). A quantitative post-mortem study of schistosomiasis mansoni in man. *American Journal of Tropical Medicine and Hygiene* 17, 38–64.

Cheever, A. W. (1985). *Schistosoma haematobium*: the pathology of experimental infection. *Experimental Parasitology* 59, 131–8.

Cheever, A., Kamel, I., Elwi, A., Moismann, J., Danner, R. & Sippel, J. E. (1978). *Schistosoma mansoni* and *S. haematobium* infections in Egypt. III. Extrahepatic pathology. *American Journal of Tropical Medicine and Hygiene* 27, 55–75.

Cline, B. L., Ryzmo, W. T., Hiatt, R. A., Knight, W. B. & Berrios-Duran, L. A. (1977). Morbidity from Schistosomiasis mansoni in a Puerto Rican community: a population-based study. *American Journal of Tropical Medicine and Hygiene* 26, 109–17.

Collins, K. J., Brotherhood, J. R., Davies, C. T. M., Dore, C., Hackett, A. J., Imms, F. J., Musgrove, J., Weiner, J. S., Amin, M. A., Awad El Karim, M. A., Ismail, H. M., Omer, A. H. S. & Sukkar, M. Y. (1976). Physiological performance and work capacity of Sudanese cane cutters with *Schistosoma mansoni* infection. *American Journal of Tropical Medicine and Hygiene* 25, 410–21.

Collins, K. J., Abdel-Rahaman, T. A. & Awad El Karim, M. A. (1988). Schistosomiasis: field studies of energy expenditure in agricultural workers in the Sudan. In: *Capacity for Work in the Tropics*, ed. K. J. Collins, & D. F. Roberts, pp. 235–47. Cambridge Univeristy Press.

Cook, J. C., Baker, S. T., Warren, K. S. & Jordan, P. (1974). A controlled study of morbidity of schistosomiasis mansoni in St. Lucian children, based on quantitative egg excretion. *American Journal of Tropical Medicine and Hygiene* 23, 625–33.

Corbett, E. L., Butterworth, A. E., Fulford, A. J. C., Ouma, J. H. & Sturrock, R. F. (1992). Nutritional status of children with schistosomiasis in two different areas of Machakos District, Kenya. *Transactions of the Royal Society of Tropical Medicine and Hygiene* 86, 266–73.

Davies, C. T. M. (1972). The effect of schistosomiasis, anaemia and malnutrition on the responses to exercise in African children. *Journal of Physiology* 230, 27P.

El Rooby, A. S., Gad El Mawla, N., Galil, N. N., Abdalla, A. & Shakkir, M. (1963). Studies on the malabsorption syndrome among Egyptians. II Malabsorption in bilharzial hepatic fibrosis. *Journal of the Egyptian Medical Association* 46, 777–82.

Farid, Z., Bassily, S., Schulert, A. R., Raasch, F., Zeind, A. S., El Rooby, A. S. & Sherif, M. (1967). Urinary blood loss in *Schistosoma haematobium* infection in Egyptian farmers. *Transactions of the Royal Society of Tropical Medicine and Hygiene* 61, 621–5.

Farid, Z., Bassily, S., Schulert, A. R., Zeind, A. S., McConnell, E. & Abdel Wahab, M. F. (1968). Urinary blood loss in *Schistosoma haematobium* infection in Egyptian farmers. *Transactions of the Royal Society of Tropical Medicine and Hygiene* 62, 496–500.

Farid, Z., Patwardhan, V. N. & Darby, W. J. (1969). Parasitism and anemia. *American Journal of Clinical Nutrition* 22, 498–503.

Farooq, M., Nielson, J., Samaan, S. A., Mallah, M. B. & Allam, A. A. (1966). The epidemiology of *Schistosoma haematobium* and *S. mansoni* infections in the Egypt-49 project area. 2. Prevalence of bilharziasis in relation to personal attributes and habits. *Bulletin of the World Health Organization* 35, 293–318.

Fenwick, A. & Figenschou, B. H. (1972). The effect of *Schistosoma mansoni* infection on the productivity of cane cutters on a sugar estate in Tanzania. *Bulletin of the World Health Organization* **47**, 567–72.

Foster, R. (1967). Schistosomiasis on an irrigated estate in East Africa. III. Effects of asymptomatic infection on health and industrial efficiency. *Journal of Tropical Medicine and Hygiene* **70**, 185.

Friis-Hansen, B. and McCullough, F. S. (1961). Anaemia and parasitic infestation in African children in Northern Rhodesia. *Journal of Tropical Medicine and Hygiene* **64**, 243–50.

Gelfand, M. (1976). The pattern of disease in Africa and the western way of life. *Tropical Doctor* **6**, 173.

Gelfand, M., Ross, M., Blair, D. & Weber, M. (1970). Schistosomiasis of the male pelvic organs. Severity of infection as determined by digestion of tissue and histologic methods in 300 cadavers. *American Journal of Tropical Medicine and Hygiene* **20**, 846.

Gerritsen, T., Walker, A. R. P., Meillon, B. & Yeo, R. M. (1953). Long term investigations of blood loss and egg load in urinary schistosomiasis in the adult African Bantu. *Transactions of the Royal Society of Tropical Medicine and Hygiene* **74**, 185–91.

Golden, D. & Barclay, R. (1972). Schistosomiasis in rural Zambia. *Annals of Tropical Medicine and Parasitology* **66**, 193–6.

Gwatkin, D. R. (1984). Does better health produce greater wealth? A review of the evidence concerning health, nutrition and output. (Unpublished USAID document.)

Holzer, B., Saladin, B., Dennis, E. & Degremont, A. (1983). The impact of schistosomiasis among rural populations in Liberia. *Acta Tropica* **40**, 239–59.

Jordan, P. & Randall, K. (1962). Bilharziasis in Tanganyika: observations on its effects and the effects of treatment in school children. *Journal of Tropical Medicine and Hygiene* **65**, 1–6.

Jordan, P. & Webbe, G. (1969). *Human Schistosomiasis*. London: Heinemann.

Kamel, I., Elwi, A., Cheever, A., Moismann, J. & Danner, R. (1978). *Schistosoma mansoni* and *S. haematobium* infections in Egypt. IV. Hepatic lesions. *American Journal of Tropical Medicine and Hygiene* **27**, 931.

Khalil, M. (1949). The national campaign for the treatment and control of schistosomiasis from the scientific and economic aspects. *Journal of the Royal Egyptian Medical Association* **31**, 817–56.

Kloetzel, K. & Da Silva, J. R. (1967). Schistosomiasis mansoni acquired in childhood: behaviour of egg counts and intradermal test. *American Journal of Tropical Medicine and Hygiene* **25**, 285–94.

Lasker, G. W. (1969). Human biological adaptability. *Science* **166**, 1480–6.

Ledward, R. (1980). Infertility in Saudi Arabia: initial experience in a new gynaecological unit. *Tropical Doctor* **10**, 117.

Lehman, J. S., Mott, K. E., Morrow, R. H., Muniz, T. M. & Boyer, M. H. (1976). The intensity and effect of infection with *Schistosoma mansoni* in a rural community in northeast Brazil. *American Journal of Tropical Medicine and Hygiene* **25**, 285–94.

Mahmood, A. (1966). Blood loss caused by helmintic infections. *Transactions of the Royal Society of Tropical Medicine and Hygiene* **60**, 766–9.

Mahmoud, A. A. F. (1980). Eosinophilopoiesis. In: *The Eosinophil in Health and*

Disease. ed. A. A. F. Mahmoud & K. F. Austen, pp. 61–75. New York: Grune & Stratton.

Mahmoud A. A. F. (1982). Schistosomiasis: clinical features and relevance to hematology. *Seminars in Hematology* **19**, 132–40.

Mansour, M. M., Francis, W. M. & Farid, Z. (1985). Prevelance of latent iron deficiency in patients with chronic *S. mansoni* infection. *Tropical and Geographical Medicine* **37**, 124–8.

McFalls, J. A. & McFalls, M. H. (1984). *Disease and Fertility.* New York: Academic Press.

Medley, G. F. & Anderson, R. M. (1985). Density-dependent fecundity in *Schistosoma mansoni* infections in man. *Transactions of the Royal Society of Tropical Medicine and Hygiene* **79**, 532–4.

Mohamed, G. H. (1992). The epidemiology of *Schistosoma mansoni* in the Gezira-Managil area of Sudan: impact of the Blue Nile Health Project. Ph.D. Thesis, University of Cambridge.

Nordbeck, H. J., Ouma, J. H. & Sloof, R. (1982). Machakos project studies. Agents affecting health of mother and child in a rural area of Kenya. XXII Schistosomiasis transmission in relation to some socio-economic and other environmental factors. *Tropical and Geographical Medicine* **34**, 193–203.

Norden, D. A. & Gelfand, M. (1973). Urinary loss of protein and amino acids in *Schistosoma haematobium* infection in West Nile, Uganda. *Transactions of the Royal Society of Tropical Medicine and Hygiene* **37**, 607–8.

Okpala, I. (1961). Studies on *Schistosoma haematobium* infection in school children in Epe, Western Nigeria. *West African Medical Journal* **10**, 401–12.

Ongom, V. L. & Bradley, D. J. (1972). The epidemiology and consequences of *Schistosoma mansoni* infection in West Nile, Uganda. *Transactions of the Royal Society of Tropical Medicine and Hygiene* **66**, 835–51.

Oomen, J. M. V., Meuwissen, J. H. H. Th. & Gemert, W. (1979). Differences in blood status of three ethnic groups inhabiting the same locality in Northern Nigeria. Anaemia, splenomegaly and associated causes. *Tropical and Geographic Medicine* **31**, 587–606.

Parker, M. (1989). The effects of *Schistosoma mansoni* on female activity patterns and infant growth in Gezira Province, Sudan. D. Phil. thesis, University of Oxford.

Pesignan, T. P., Hariston, N. G., Jauregui, J. J., Garcia, E. G., Santos, B. C. & Besa, A. A. (1958). Studies on *Schistosoma japonicum* infection in the Philippines. 1. General considerations and epidemiology. *Bulletin of the World Health Organization* **18**, 465–75.

Prescott, N. M. (1979). Schistosomiasis and development. *World Development* **7**, 1–14.

Pugh, R. N. H. & Gilles, H. M. (1978). Malumfashi endemic diseases research project, III. Urinary schistosomiasis: a longitudinal study. *Annals of Tropical Medicine and Parasitology* **37**, 170–1.

Retel-Laurentin, A. (1978). Appraising the role of certain diseases in sterility. *Population* **33**, 101.

Salih, S., Marshall, T. F. de C. & Radalowicz, A. (1979). Morbidity in relation to the clinical forms and to intensity of infection in *Schistosoma mansoni* infections in the Sudan. *Annals of Tropical Medicine and Parasitology* **73**, 439–49.

Sazazuki, T. Ohuta, N., Kanoeoka, R. & Kojima, S. (1979). Association between an HLA haplotype and low responsiveness to schistosomal worm antigen in man. *Journal of Experimental Medicine* **152**, 314–8.

Scott, D., Senker, K. & England, E. C. (1982). Epidemiology of human *Schistosoma haematobium* infection around Volta Lake, Ghana, 1973-1975. *Bulletin of The World Health Organization* **60**, 89–100.

Stephenson, L. S. (1987). *Impact of Helminth Infections on Human Nutrition.* London: Taylor & Francis.

Stephenson, L. S., Latham, M. C., Kinoti, S. N. & Oduori, M. L. (1985a). Regression of splenomegaly and hepatomegaly in children treated for *Schistosoma haematobium* infection. *American Journal of Tropical Medicine and Hygiene* **34**, 119–23.

Stephenson, L. S., Latham, M. C., Kurz, K. M., Kinoti, S. N., Oduori, M. L. & Crompton, D. W. T. (1985b). Relationships of *Schistosoma haematobium*, hookworm and malarial infections and metrifonate treatment to growth of Kenyan school children. *American Journal of Tropical Medicine and Hygiene* **34**, 1109–18.

Stephenson, L. S., Latham, M. C., Kurz, K. M., Kinoti, S. N., Oduori, M. L. & Crompton, D. W. T. (1985c). Relationships of *Schistosoma haematobium*, hookworm and malarial infections and metrifonate treatment to haemoglobin level in Kenyan school children. *American Journal of Tropical Medicine and Hygiene* **34**, 519–28.

Stephenson, L. S., Latham, M. C., Kurz, K. M., Miller, D., Kinoti, S. N. & Oduori, M. L. (1985d). Urinary iron loss and physical fitness of Kenyan children with urinary schistosomiasis. *American Journal of Tropical Medicine and Hygiene* **34**, 322–30.

Stephenson, L. S., Latham, M. C., Kurz, K. M., Miller, D. & Kinoti, S. N. (1986). Relationships of *Schistosoma haematobium*, hookworm, and malarial infections and metrifonate treatment to nutritional status of Kenyan coastal school children: a 16 month follow-up. In: *Schistosomiasis and Malnutrition*, ed. L. S. Stephenson, pp. 27–65. Cornell International Nutrition Monograph Series no. 16. Ithaca, New York: Cornell International Nutrition Program.

Van Ee, J. H. & Polderman, A. M. (1984). Physiological performance and work capacity of tin mine labourers infested with schistosomiasis in Zaire. *Tropical Geographical Medicine* **36**, 259–66.

Von Lichtenberg, F. (1987). Consequences of infections with schistosomes. In: *The Biology of Schistosomes*, ed. D. Rollinson, & A. J. G. Simpson, pp. 185–232. London: Academic Press.

Walker, A. R. P., Faith Walker, B. & Richardson, B. D. (1970). Studies on schistosomiasis in a South African Bantu population. *American Journal of Tropical Medicine and Hygiene* **19**, 792–814.

Walker, A. R. P., Faith Walker, B., Richardson, B. D. & Smit, P. J. (1972). Running performance in South African Bantu children with schistosomiasis. *Tropical and Geographical Medicine* **24**, 347–52.

Walsh, J. A. (1990). Estimating the burden of illness in the Tropics. In: *Tropical and Geographical Medicine*, ed. K. S. Warren & A. A. F. Mahmoud, pp. 185–96. New York: McGraw-Hill.

Warren, K. S. (1982). The secret of the immunopathogenesis of schistosomiasis: in vivo models. *Immunological Reviews* **61**, 189–213.

Waslein, C. I., Farid, Z. & Darby, W. J. (1973). The malnutrition of parasitism in Egypt. *Southern Medical Journal* **66**, 47–50.

Weisbrod, B. A., Andreano, R. L., Baldwin, R. E., Epstein, E. M. & Kelley, A. C. (1973). *Disease and Economic Development. The impact of parasitic diseases in St. Lucia.* University of Wisconsin Press.

Weisbrod, B. A., Andreano, R. L., Baldwin, R. E., Epstein, E. M. & Kelley, A. C. (1974). Disease and economic development. *International Journal of Social Econometrics* **1**, 111–17.

Wilkins, H. A., Blumenthal, U. J., Hagan, P., Hodgson, J. & Menon, A. (1986). Transmission of urinary schistosomiasis in man-made habitats in the Gambia. *Transactions of the Royal Society of Tropical Medicine and Hygiene* **80**, 1009.

Wilkins, H. A., Goll, P., Marshall, T. F. de C. & Moore, P. (1979). The significance of proteinuria and hematuria in *Schistosoma haematobium* infection. *Transactions of the Royal Society of Tropical Medicine and Hygiene* **73**, 74–80.

World Health Organization (1981). *International Classification of Diseases, Injuries, and Causes of Death.* Geneva: WHO.

World Health Organization (1985). *The control of schistosomiasis.* (Report of WHO expert committee.) Technical Report Series no. 728. Geneva: WHO.

Wright, W. H. (1972). A consideration of the economic impact of schistosomiasis. *Bulletin of the World Health Organization* **47**, 559–66.

10 Biological adaptability, plasticity and disease: patterns in modernizing societies

RALPH M. GARRUTO

Summary

Human population biology is one of the few remaining integrative multidisciplinary sciences in today's reductionist scientific world. The reductionist approach is perhaps nowhere better exemplified than in the biomedical sciences where both long-term research and the ability to integrate and synthesize large bodies of data toward a solution of a broadly defined problem are now all but non-existent. In contrast, the concepts of human adaptability and plasticity and their development within an ecological framework emerged from the so-called 'golden age' of human biology in the 1960s and continue to thrive despite the changes that have taken place over the past three decades. This theme focuses on issues of human–environment interactions and resultant biological and behavioral outcomes in populations worldwide. The golden age of human biology flourished at a time when research dollars were plentiful, theoretical conceptualizations and constructs broad-based, multidisciplinary approaches emphasized and when many isolated human groups were beginning to emerge into the modern world. The historical and theoretical aspects regarding biological adaptability studies are appropriately addressed by L. Schell (this volume) and will not be reiterated here.

Among the contributors and supporters to an understanding of human biological adaptability was Professor Gabriel Lasker, whose 1969 article (*Science* 166, 1480–6) helped focus this issue by presenting a three-part construct of human biological adaptation that was defined as '. . . a modification in structure or function that enables an organism to survive and reproduce' (Lasker, 1969). It was represented by a temporal sequence that included the concepts of genetic adaptation, developmental adaptation or plasticity (lifelong) and acclimatization (short-term). The terms adaptation and plasticity were used not only for entire populations or groups but also for individuals, organ systems and cellular populations, although the application of the construct by human biologists was usually defined within the context of entire human populations.

Summarized below is an attempt to demonstrate the importance of integrative research using state-of-the-art laboratory methods and cross-disciplinary field research designs to answer major biomedical problems and inquiries. I describe research examples from our long-term field and laboratory work that represent not only the three points in the temporal continuum (i.e. genetic adaptation, developmental adaptation or plasticity and acclimatization) in modernizing human populations, but also examples at the individual and cellular levels through *in vivo* and *in vitro* experimental models that are a direct outgrowth of our human population studies. Finally, I elaborate on new models in nature that link us back to our initial studies of human disease and the concepts of adaptability and plasticity.

Gene–environment interactions at high altitude in Qinghai, China

Living at high altitude poses a number of severe stresses for human populations. The most significant is the low partial pressure of oxygen, which demands both morphological and physiological compensations to insure that tissues have an adequate supply of oxygen. Our current research among high-altitude peoples living in Qinghai is a direct outgrowth of our earlier work on the adaptation of human populations to hypoxic stress in the Andes more than a quarter-century earlier, during the so-called 'golden age' of human biology (Baker, 1966, 1969; Garruto & Dutt, 1983; Frisancho & Greksa, 1989; Ballew, Garruto & Haas, 1989). Then, as now, one of the most interesting and important theoretical questions is whether human populations with distinct gene pools adapt similarly to a given unmodifiable environmental stress and whether such adaptations are short-term (acclimatization) developmental (lifelong-biological plasticity) or generational (genetic adaptation). The conventional wisdom, based on many studies conducted in the Andes, is that indigenous high-altitude peoples possess large lungs to maximize oxygen extraction and high numbers of red blood cells to maximize oxygen transport to the tissues. But recent research has suggested that Asian highlanders possess smaller chests and lower lung function values relative to body size and lower hemoglobin concentrations and red blood cell counts compared with Andeans (Beall, 1982, 1984; Beall *et al.*, 1990). Indeed, the pulmonary function and hemoglobin values of Asian highlanders may be within the ranges of normal, sea-level populations, possibly indicating that indigenous Himalayan and Tibetan populations possess a superior, genetically determined physiological adjustment to hypoxia, which reduces the need for compensatory lung and hematological responses (e.g. Beall *et al.*, 1990; Winslow, Chapman & Monge, 1990). However, only a few systematic studies have been conducted at high altitude in Asia, and not all results fit the hypothesis.

Table 10.1. *High-altitude research design for the study of gene–environment interactions in Qinghai Province, People's Republic of China*

The table shows the stratification of samples by age, sex, village, altitude, and indigenous or migrant status.

Village (permanent)	Sex/age (permanent)	Sample Sizes			
		Han (permanent)	Tibetan (migrants)	Hui	Han
Mado (4300 m)	Boys	100	100	100	100
	Girls	100	100	100	100
	Male adults	NA [a]	40	40	40
	Female adults	NA	40	40	40
Guinan (3200 m)	Boys	150	150	150	150
	Girls	150	150	150	150
	Male adults	40	40	40	40
	Female adults	40	40	40	40
Beijing (sea level)	Boys	200	NA	200	—
	Girls	200	NA	200	—
	Male adults	40	NA	40	—
	Female adults	40	NA	40	—

[a]NA, not available.

In the high-altitude province of Qinghai in western China an opportunity exists to test the hypothesis that Tibetan populations possess thorax dimensions, pulmonary function and hematological characteristics that are similar to those of lowlanders, and to examine gene–environment interactions among different populations (Garruto *et al.*, 1991; Shen *et al.*, 1992; Weitz *et al.*, 1992; Weitz & Garruto, 1993). Qinghai Province has a total population of slightly over 4 million and a land mass of 721 million km^2, of which 54% lies between 4000 m and 5000 m, with an additional 24% between 3000 m and 4000 m (Fig. 10.1). The inhabitants include several linguistically and genetically different populations that are indigenous to this high-altitude region, as well as first- and second-generation migrant Han people of low-altitude ancestry. Qinghai is the only location in Asia where indigenous and genetically diverse highland groups live in close proximity to large populations of Han who have been born and raised at high altitude.

The sampling strategy summarized in Table 10.1 is designed to compare the characteristics of children, adolescents and adults from the three primary groups inhabiting the region: indigenous Golak Tibetans who have inhabited this region for millennia, indigenous Hui (Chinese Muslims)

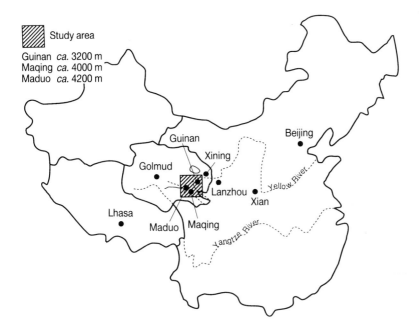

Fig. 10.1. Map of the study area for gene–environmental interaction at high altitude in Qinghai Province, People's Republic of China.

who have inhabited the area since the eighth and ninth centuries and who have not intermarried with other groups (Ekvall, 1939), and individuals of Han descent who were born and raised at high altitude. Our study takes a developmental approach to understanding the physiological and anatomical characteristics that occur among adults. It is based on the knowledge that children and adolescents have a greater potential to make significant adjustments to environmental stress than adults, regardless of genetic background. Finally, it is designed to determine whether population differences exist in the *potential* to adjust to hypoxia during development. If physiological and anatomical responses to hypoxia are related to altitude effects alone, they should develop in a similar way during child growth and development in all groups regardless of ethnic background. However, if physiological and anatomical responses to hypoxia are determined genetically or as the result of different gene–environment interaction, development should differ among populations with similar exposure to hypoxia. Our study controls for environmental conditions by studying only children who have been exposed to high altitude since birth, by comparing children who have been born and raised at the same altitude, by comparing children who have not visited lower or higher elevations in the year preceding the study, and by comparing children of similar socioeconomic backgrounds, all of whom are in good health and well-nourished.

Studies comparing the growth of children from genetically different populations born and raised at the same high altitude have worked effectively in the Andes (Greksa, 1990). In his study of indigenous Aymara boys and boys of European descent who were born and raised at high altitude in Bolivia, Greksa (1986) reported that, unlike the Peruvian Quechua, neither group showed a particular 'spurt' in chest growth or pulmonary function during adolescence. However, he determined that thorax and pulmonary function growth in both groups was accelerated over Europeans born at low altitude (Greksa, 1986, 1987, 1988; Greksa & Beall, 1989). Still, Aymara boys had larger chest measurements relative to stature than boys of European descent who were born and raised at high altitude (Greksa, 1986, Greksa & Beall, 1989). Thus, although both groups exhibited a clear altitude-related response, the response potential appeared to be greater among the indigenous Aymara.

Our preliminary observations in Qinghai indicate that Tibetan, Han and Hui children between the ages of 6 and 12 years, who were born and raised at high altitude, show no significant differences in chest dimensions or pulmonary function values; but these children do show larger chests and higher pulmonary function values than predicted for children of similar ages and statures at low altitude. Beginning at adolescence, however, Tibetan and Hui groups between the ages of 13 and 19 years show a more rapid increase in thorax dimensions (particularly chest depth) relative to stature than occurs among adolescent Han. A similar divergence also occurs in pulmonary function values. When compared with Peruvian Quechua, Han, Hui and Tibetan children and adolescents of all ages in Qinghai have smaller thoraxes and lower pulmonary function values.

Preliminary observations of hematological characteristics among Han, Hui and Tibetan children born and raised at the same altitude in Qinghai suggest that they are virtually identical throughout childhood and adolescence. During adolescence, the variability in hemoglobin concentration [Hb] and hematocrit increases significantly in all groups, with a surprising number of Han and Tibetan adolescents showing significantly elevated values. We also find that some adolescents appear to experience hypoxemia; and we have reported the existence of a negative correlation between arterial oxygen saturation and [Hb] among Han and Tibetan adolescents at both 3200 m and 4300 m (Weitz & Garruto, 1993).

By extending our current research design in Qinghai to include Tibetan nomads, as well as settled populations, and by developing a semilongitudinal approach to investigate the link between hematology, hypoxemia and ventilatory response, we hope to answer major questions about the biology and mechanisms of Asian adaptability to hypoxia.

Nutritional stress and concepts of optimal and critical nutrient levels

Perhaps no single factor influences biological outcomes and patterns of disease as much as nutrient availability and nutritional practices of the world's populations. It is also clear that nutrition is one of the most difficult factors to evaluate, particulary at a secondary level of association with disease, and is affected by group mobility, uniformity and availability of food sources, specific cultural practices and ecosystem stability. One of the best and seemingly least complicated examples of nutritional stress leading to frank clinical disease is that of simple iodine deficiency and the development of hyperendemic goiter, cretinism and associated deaf mutism among the Dani people of West New Guinea. Iodine deficiency is a serious problem, even today, in many inland mountain regions of the world and leads to a crippling load of defect in isolated as well as in modernizing societies that live in iodine-deficient areas. The work of our laboratory in this area began more than three decades ago in the central highlands of West New Guinea with descriptions of the problem by Gajdusek in the upper Ruffaer River around the Mulia Mission Station in the western Dani linguistic group (Van Rhijn, 1960; Gajdusek, 1962; Kidson & Gajdusek, 1962) (Fig. 10.2). Subsequent studies further delineated the extent of this focus and attempted to predict clinical outcomes on the basis of serum protein-bound iodine (PBI) concentrations (Curtain *et al.*, 1974; Gajdusek, Garruto & Dedecker, 1974; Garruto, Gajdusek & ten Brink, 1974; Gajdusek & Garruto, 1975).

Within a single cultural and linguistic group, goiter may attain a very high incidence in one region and a very low incidence in an adjacent region, without any change in diet, other cultural factors, language, or genetic origin. Thus, the western Dani in three different neighboring locations show three very different patterns of goiter incidence (Fig. 10.2). In the Yembi–Mulia–Guderi valleys, over 50% of all adult women from most villages suffer from goiters, many of very large size; in the Mamit and Karubaga areas of the Swart Valley goiter is low (less than 5%); and in the Ilu region goiter is absent, except when introduced in a rare immigrant. Similarly, the Uhunduni people of Illaga, Nuema, and Beoga suffer from goiter at rare, low, and high incidences, respectively.

Clinical goiter in these populations is usually associated with a spectrum

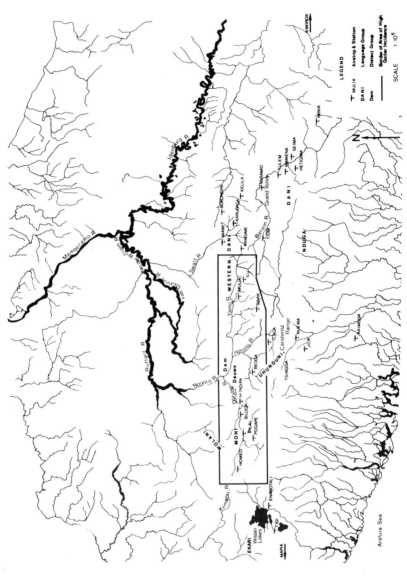

Fig. 10.2. Map of the Central Highlands of West New Guinea showing the belt of hyperendemic goiter, cretinism and associated deaf-mutism and adjacent regions and villages of low goiter incidence.

of neurological and skeletal defects that include classical cretinous dwarfism, deaf–mutism, motor deficits, and varying degrees of mental subnormalities. Iodine deficiency is clearly the cause of endemic goiter and cretinism in these inland populations and endemic cretinism results from fetal damage *in utero* due to iodine deficiency of the mother during pregnancy. Indeed, the incidence of children of all ages with severe mental or neuromotor deficits or deaf–mutism is over 12%, and reaches 20% in some villages.

Our studies demonstrated that populations of the region with low-incidence goiter have no higher serum protein-bound iodine concentrations than do those groups who are severely affected. We concluded that serum iodine determinations do not serve to indicate the severity of the goiter problem in a given region (Garruto *et al.*, 1974). Thus, we are faced with the enigma of why we cannot predict clinical outcomes in different villages or even within the same village among people who speak the same language, have the same cultural practices and seemingly the same diet and nutritional status, and who share the same gene pool.

Given the above circumstances, it is useful to invoke the concepts of optimal and critical levels in populations precariously adapted to their environment. In areas where extremely low levels of iodine exist, populations may be 'pregoitrous' and adapted to low 'critical' levels of iodine intake. However, the slightest decrease or shift in iodine intake and availability may provoke marginal-risk individuals into full-blown clinical disease with a wide spectrum of defects. The concept of biological plasticity allows for adaptation to low nutrient requirements such as iodine, even optimally so, but in the face of populations precariously adapted to biologically critical levels of iodine availability the situation becomes unpredictable. Thus, we have struggled with the idea of how goiter has newly appeared in some villages in this region that have remained sedentary and has slowly disappeared in others without noticeable alterations in diet. We concluded that in analogy to the goiterogenic action of cabbage, that in a region where sweet potato tubers (the New Guinea staple food) do not grow well, their replacement by large quantities of iodine-binding sweet potato leaves in the diet in areas of marginal iodine availability may be the 'straw that breaks the camel's back'. The small amount of free circulating iodine in an individual is bound by the goiterogenic action of the sweet potato leaves and this once available iodine source is eventually lost in the stool. With the advent of increasing modernization and contact in the 1960s, iodized oil injections given to the population acted as a simple, inexpensive, long-term therapy to completely resolve the problem and restore the health of the population.

Neuronal degeneration in Pacific populations and the development of adaptive strategies at the population, individual and cellular levels

During the past three decades, the opportunistic and multidisciplinary study of hyperendemic foci of amyotrophic lateral sclerosis (ALS) and parkinsonism–dementia (PD), which occur in different cultures, in different ecological zones and among genetically divergent populations, have served as natural models that have had a major impact on our thinking and enhanced our understanding of these and other neurodegenerative disorders such as Alzheimer's disease and the process of early neuronal aging. Our cross-disciplinary approach to these intriguing neurobiological problems and the accumulated epidemiological, genetic, cellular and molecular evidence strongly implicates environmental factors in their causation, specifically the role of aluminum and its interaction with calcium in neuronal degeneration (Garruto, 1991).

Amyotrophic lateral sclerosis (ALS), also known as Lou Gehrig's disease, is a disorder affecting the motor neurons of the brain's motor cortex, brain stem and spinal cord and the neural pathways between them (corticospinal tracks). The resulting clinical symptoms include progressive weakness and atrophy of skeletal muscle, particularly noticeable in the limbs and shoulder girdle, and difficulty with speech and in swallowing. The disease is relentlessly progressive and leads to near-total paralysis and death, generally within four years.

Parkinsonism–dementia (PD), a second neurological disorder found together in high incidence with ALS in the same population, same sibship and occasionally in the same individual, results in profound mental deterioration from degeneration of neurons in the cerebral cortex and midbrain. The progressive dementia is accompanied by parkinsonian features of slowness of voluntary motor activity, muscular rigidity, tremor and blank facial masking (staring gaze). It, too, is uniformly fatal with a course usually lasting 4–5 years.

Both disorders are found to occur hyperendemically among the Chamorro people of Guam and Rota in the Mariana Islands (Kurland & Mulder, 1954; Garruto, 1991), among the Auyu and Jakai people of southern West New Guinea (Gajdusek & Salazar, 1982) and among Japanese from the Kii Peninsula of Japan (Shiraki & Yase, 1975) (Fig. 10.3). On Guam, ALS and PD originally occurred at incidence rates 50–100 times higher than that found elsewhere in the world, with 1 in 5 people over age 25 dying from these disorders, making it a major public health problem. Similar rates are found in the Kii Peninsula focus, and in West New Guinea the rates are even higher. However, over the past 30 years there has been a dramatic

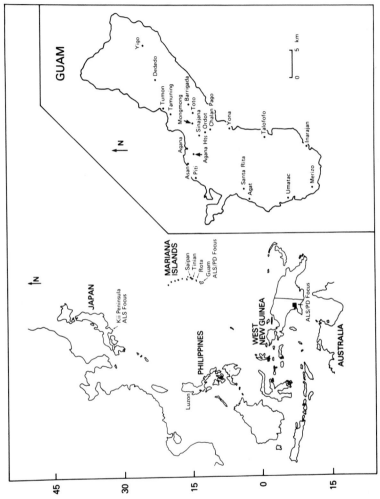

Fig. 10.3. Map showing the location of the three geographically, culturally, and genetically distinct foci of amyotrophic lateral sclerosis (ALS) and parkinsonism–dementia (PD) in the western Pacific. These uniformly fatal disorders affect the Chamorro people of Guam and Rota in the Mariana Islands, the Auyu and Jakai people of southern West New Guinea, and Japanese from the Kii Peninsula of Japan.

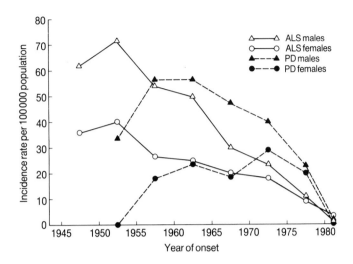

Fig. 10.4. Five-year average annual incidence rates of amyotrophic lateral sclerosis (ALS) and parkinsonism–dementia (PD) in males and females by year of onset. The data are age- and sex-adjusted to the 1960 Guamanian Chamorro population.

decline in incidence in all three foci in the western Pacific with increasing modernization and contact and introduction of new foodstuffs, supporting the idea of a cohort effect and an environmental cause for these disorders (Fig. 10.4) (Garruto, Yanagihara & Gajdusek, 1985; Garruto, 1991). Thus, we have proposed as a working hypothesis that a basic defect in mineral metabolism induces a form of secondary hyperparathyroidism, provoked by chronic nutritional deficiencies of calcium, leading to enhanced gastrointestinal absorption of bioavailable aluminum and the deposition of calcium and aluminium in neurofibrillary tangle-bearing neurons, the hallmark neuropathological lesion in these disorders.

In more than three decades of searching, all studies have failed to indicate any satisfactory genetic explanation for ALS and PD (Garruto, 1991), including the recent discovery in familial ALS of mutations in the superoxide dismutase 1 (SOD1) gene (Rosen *et al.*, 1993). Our current search for such mutations using DNA from ALS and PD patients thus far has been negative for any point mutations in the 5 exon regions of the SOD1 gene (Figlewicz *et al.*, 1994). Point mutations in the exon regions of the SOD2 gene are also under investigation in a continuing search to exclude a major gene effect.

Pathologically, there is a dramatic loss of neurons in brain and spinal cord of these patients, accompanied by a number of other neuropathological

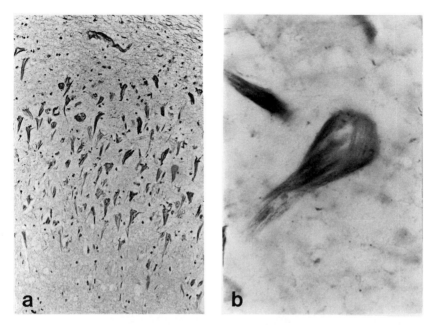

Fig. 10.5. (*a*) Histochemical silver staining (Bodian stain) to demonstrate extensive numbers of neurofibrillary tangles (NFT) found in various subanatomic regions of the brain and spinal cord of patients with amyotrophic lateral sclerosis (ALS) and parkinsonism–dementia (PD). The NFTs are a hallmark neuropathological characteristic of these disorders. Region shown is the pyramidal cell layer of the hippocampal region of the brain in a patient with PD. (*b*) high-power view of a NFT in the hippocampus from the same area (different field) demonstrating the fibrillary nature of the proteins that accumulate.

changes including the hallmark lesion of neurofibrillary tangles in cortical and spinal neurons (Fig. 10.5). A number of years ago we demonstrated the colocalization of calcium and aluminum in these neurofibrillary tangle-bearing neurons (Fig. 10.6) (Garruto *et al.*, 1984). It is likely that the interaction of these elements is causally involved in the disease process and that there is considerable biological variability in response to these toxic events as indicated by 70% of the normal Chamorro population developing neurofibrillary tangles over age 25 (preclinical state) although only 20% actually develop clinical disease. The differential toxicity is probably due to factors such as aluminum speciation and bioavailability, a reduced protective mechanism in the gut and/or kidney, a compromised blood–brain barrier, and altered cell membrane permeability that leads to the manifestation of clinical disease in some individuals and the development of neurofibrillary tangles. Changes in the dietary intake of calcium and bioavailable aluminum

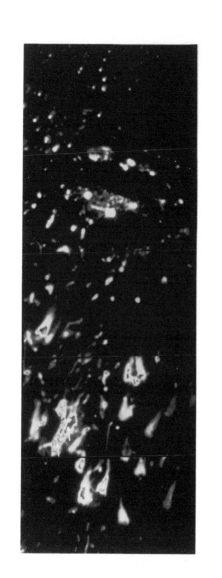 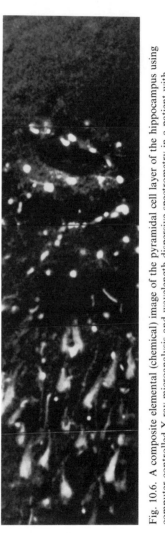

Fig. 10.6. A composite elemental (chemical) image of the pyramidal cell layer of the hippocampus using computer controlled X-ray microanalysis and wavelength-dispersive spectrometry in a patient with parkinsonism–dementia. These elemental images show the striking colocalization of calcium (top) and aluminum (bottom) in the same neurofibrillary tangle-bearing neurons. Semiquantitative estimates for the brightest images in the field are 7200 ppm and 500 ppm for calcium and aluminum, respectively (for analytical details and methodology see Garruto et al., 1984).

and the subsequent correction of any secondary hyperparathyroidism would likely break the cycle of high-incidence disease, but would be unlikely to prevent or reverse the neuronal damage and cell death that these deposits have already produced. The eventual outcome to these biomedical riddles awaits a resolution of the problem through the development of appropriate experimental models and designs.

Experimental studies demonstrating individual and cellular levels of adaptation

As a direct consequence of our studies in these Pacific populations, we have undertaken the long-term development of experimental models of neuronal degeneration, in an attempt to understand the cellular and molecular mechanisms by which toxic metals affect the central nervous system (Garruto, Flaten & Wakayama, 1993). Our experimental studies have resulted in the establishment of an aluminum-induced chronic myelopathy in rabbits (Strong *et al.*, 1991) and the development of neurofibrillary tangle-like lesions after low-dose aluminum administration in cell culture (Strong & Garruto, 1989, 1991). These studies clearly demonstrate the philosophy that chronic rather than acute experimental models of toxicity are necessary in order to enhance our understanding of human neuro-degenerative disorders with long latency and slow progression. Likewise, it demonstrates quite well the concepts of individual and cellular levels of biological variation, plasticity and adaptation described by Lasker more than 20 years ago (Lasker, 1969).

In the following example, I illustrate how our research directions changed from studying human populations to studying the individual organism and cellular populations as a means of trying to understand the pathogenic mechanism involved in ALS and PD (Fig. 10.7). This research over the past several years has involved extensive experimental modeling both *in vivo* and *in vitro* and has yielded some rather fascinating results, not only with respect to pathogenic mechanisms, but also with regard to individual variation and biological plasticity of cellular populations. It gives us some insight into the kinds of variability that may occur in human diseases at the population level.

In the first example, a model of chronic intoxication was developed by administering a low dose (100 μg) of aluminum chloride ($AlCl_3$) intracis-ternally in New Zealand white rabbits once monthly (Strong *et al.*, 1991). The animals were examined neurologically at least once a week over a course of 8–9 months. Fig. 10.8 shows the weekly clinical score from a single rabbit litter (aged 6 weeks at the start of the experiment) plotted against time (the higher the score, the more severe the toxic effect of aluminum). As

ADAPTATION and PLASTICITY

'........ modification in structure or function that
enables an organism to survive and reproduce.'
(Lasker, 1969)

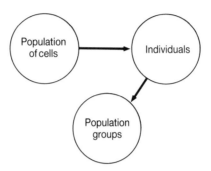

Fig. 10.7. Simplified cartoon illustrating the concepts of adaptation and
plasticity at different levels of integration (cellular, individual and population).

can be seen there is enormous variability in response to aluminum with
respect to the development of clinical signs using the same procedure, same
route of inoculation, and same aluminum dose during the course of the
8-month study. Some littermates became very toxic and demonstrated
severe clinical signs over a short period of time to this low, once monthly
dose of aluminum, whereas others responded very slowly to the same dose.
These data demonstrate well the concept of individual variability and
plasticity in a mammal population not unlike that seen in human ALS and
PD and in goiter and cretinism in West New Guinea.

Using a second example to further demonstrate the concept of biological
variability and plasticity, but at the cellular level, rabbit hippocampal and
motor neurons were grown in a chemically defined medium and tested for
their response to a single low dose of $A1C1_3$ at 1, 10, 25, 50 and 100 μM
concentrations (Strong & Garruto, 1989, 1991). Our observations demonstrate that spinal motor neurons had an order of magnitude greater
sensitivity to aluminum toxicity than did hippocampal neurons cultured
under identical conditions and given the same dose of aluminum. Both of
the above examples make the point that biological variation and plasticity
occur at different levels (i.e. population, individual, cellular) and in different
contexts, but all can be combined in a multidisciplinary theoretical
framework to help answer major biomedical questions of worldwide
importance.

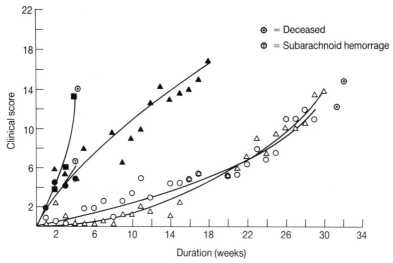

Fig. 10.8. Plasticity of responses to chronic low-dose aluminum intoxication in New Zealand white rabbits inoculated intracisternally once monthly with 100 μg AlCl₃. All rabbits are from a single litter and developed progressive hyperreflexia, hypertonia, splaying of the limbs and gait abnormalities at different rates. Some rabbits became more acutely ill than their littermates; others were much more slowly affected by the aluminum. This biological variability within the same litter has direct relevance to individual variability and biological plasticity in humans. (For details see Strong *et al.*, 1991.)

Future research directions and new lessons in nature

In an attempt to come full-circle with our work on aluminum intoxication, I discuss a current research effort that takes advantage of newly discovered models in nature as an outgrowth of both our human and experimental studies. This effort involves a natural model of aluminum intoxication in fish from acid-rain lakes (Fig. 10.9). We know that acid rain leads to the leaching of aluminum from soils deficient in calcium minerals (which act as a buffer), and fish exposed to lakes and streams with high concentrations of dissolved aluminum die of aluminum intoxication (Driscoll *et al.*, 1980). The primary target organ in mature fish dying of aluminum intoxication is the gills, where aluminum interferes with iono- and osmoregulation, leading to respiratory dysfunction and death. However, such natural models have not been exploited for what they might tell us about the chronic effects of aluminum on the nervous system, the mechanisms of aluminum intoxication and the identification of bioactive forms of aluminum, be they mononuclear or polynuclear species.

We are now in the process of initiating new research in our long-term

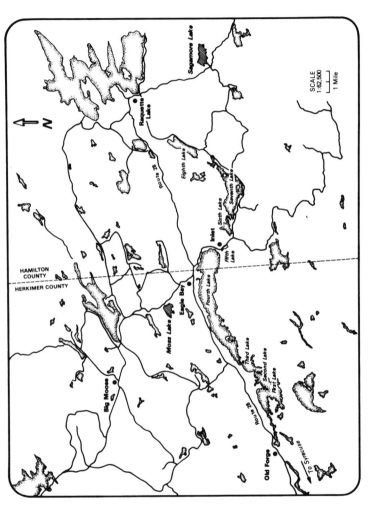

Fig. 10.9. Map of the study area in the western Adirondack Mountains of New York State, an area heavily affected by acid rain. Various species of fish were obtained from lakes with a variety of water chemistries and aluminum concentrations. Eighth Lake is a control lake with a neutral pH and very low aluminum concentrations. Moss Lake is distinctly acid with elevated aluminum levels, but an interactive chemistry between aluminum and dissolved organic carbon in the lake renders it less harmful to fish populations. Big Moose and Sagamore Lakes have high aluminum

Table 10.2. *Number of species of fish from lakes in acid rain study areas in the western Adirondack Mountains of New York prepared for neuro- pathological and elemental analysis*

Species	Fish hatchery No. (age)	Eighth Lake No. (age)	Moss Lake No. (age)	Sagamore Lake No. (age)	Big Moose Lake No. (age)
Rainbow trout *Oncorhynchus mykiss*	5 (1,5,2.5 yrs)—	—	—	—	—
Brook trout *Salvelinus fontinalis*	—	—	1 (3 yrs)	5 (2–4 yrs)	—
Lake trout *Salvelinus namaycush*	—	5 (5–6 yrs)	—	2 (6,11 yrs)	—
White sucker *Catostomus commersoni*	—	2 (4 yrs)	3 (3,5,8 yrs)	1 (5 yrs)	8 (3–4 yrs)
Longnose sucker *Catostomus catostomus*	—	—	—	2 (6 yrs)	—
Yellow perch *Perca flavescens*	—	—	3 (3,7,8 yrs)	—	5 (3–7 yrs)
Rock bass *Ambloplites rupestris*	—	3 (3–6 yrs)	—	—	—
Brown bullhead *Ictalurus nebulosus*	—	2 (no age)	—	—	—

effort to understand the basic cellular and molecular mechanisms of aluminum intoxication and metal–metal, metal–gene and metal–enzyme interactions in the central nervous system. In collaboration with wildlife biologists from the New York State Department of Environmental Conservation, we captured and perfused, as well as froze, tissue from a number of species of fish from acid-rain lakes in the Adirondack Mountains of New York. Preliminary results suggest that the central nervous system of these fish have neuropathological changes consistent with those found in human ALS and in experimental aluminum intoxication (Flaten *et al.*, 1993). Our studies have also demonstrated the focal deposition of aluminum in the sensory cell layer of the olfactory epithelium and the diffuse accumulation of aluminum in the olfactory glomeruli layer of the olfactory bulb. We are currently looking at different species of fish under the same and different water chemistries in order to assess different levels and mechanisms of adaptation and species-specific differences in aluminum toxicity (Table 10.2). Also, it has not gone unnoticed by us that patients with Alzheimer's disease (Harrison, 1990), parkinsonism (Doty, Deems & Stellar, 1988), motor neuron disease (Elian, 1991), and recently those with Pacific ALS and PD (Doty *et al.*, 1991) have been found to have olfactory deficits and neurofibrillary tangles in the olfactory cortex, and that environmental agents such as aluminum, in either mononuclear or polynuclear chemical forms (Flaten & Garruto, 1992), have direct access to the brain from the olfactory neural pathway.

The toxic effects of aluminum on fish as a result of acid-rain pollution offer several exciting new models to explore. These natural fish models are chronic (in non-migrating species) as well as acute. Yet, the models also can be chronic–intermittent, as with offspring of some salmon species that hatch, develop, and mature in acid-rain lakes, rivers and streams with high concentrations of aluminum, only to migrate to the sea to start the salt-water phase of their life cycle, subsequently returning again some years later to spawn in the exact same acid river or stream in which they were born. The experimental possibilities and potential insights from just this one example are many and combine laboratory and field approaches.

In summary, I wish to emphasize that seldom do we have the opportunity to explore in-depth hyperendemic foci of neurodegenerative disorders, often with long latency and slow progression, and that such natural paradigms greatly extend our research potential and offer unparalleled opportunities for understanding disease etiology and mechanisms of pathogenesis. Secondly, the concepts of biological variation, adaptation and plasticity can be studied with great success at the population, individual, and cellular levels using long-term, highly integrated, multidisciplinary approaches involving natural and experimental models in both field and laboratory settings. Finally, it is clear that a broad-based integrative approach (human biology) and a reductionist approach (biomedicine) are not necessarily incompatible or mutually exclusive. Indeed, in this chapter examples of both approaches have been successfully used toward the solution of major scientific questions. In the end it is the framing of the question that is important; it allows us to put into place an appropriate research design that will eventually generate the necessary data to help answer the question at hand. In an ideal world this scientific process is likely to lead to the development of broad generalizations about natural phenomena and the raising of new questions for exploratory investigation. Human biology's greatest asset is the ability of its disciples to plan, conduct and synthesize broad-based research toward the solution of major scientific questions.

References

Baker, P. T. (1966). Human biological variation as an adaptive response to the environment. *Eugenics Quarterly* **13**, 81–9.

Baker, P. T. (1969). Human adaptation to high altitude. *Science* **163**, 1149–56.

Ballew, C. L., Garruto, R. M. & Haas, J. D. (1989). High altitude hematology: paradigm or enigma? In: *Human Population Biology: A Transdisciplinary Science*, ed. M. A. Little & J. D. Haas, pp. 239–62. New York: Oxford University Press.

Beall, C. M. (1982). A comparison of chest morphology in high altitude Asian and Andean populations. *Human Biology* 54, 145–63.

Beall, C. M. (1984). Aging and growth at high altitude in the Himalayas. In: *The People of South Asia*, ed. J. R. Lukacs, pp. 365–85. New York: Plenum Press.

Beall, C. M., Brittenham, G. M., Macuaga, F. & Barragan, M. (1990). Variation in hemoglobin concentration among samples of high-altitude natives in the Andes and the Himalayas. *American Journal of Human Biology* **2**, 639–51.

Curtain, C. C., Gajdusek, D. C., O'Brien, D. & Garruto, R. M. (1974). Congenital defects of the central nervous system associated with hyperendemic goiter in a neolithic highland society of Western New Guinea, IV: Serum proteins and haptoglobins, transferrins and hemoglobin types in goitrous and adjacent non-goitrous Western Dani. *Human Biology* **46**, 331–8.

Doty, R. L., Deems, D. A. & Stellar, S. (1988). Olfactory dysfunction in parkinsonism: a general deficit unrelated to neurologic signs, disease stage, or disease duration. *Neurology* **38**, 1237–44.

Doty, R. L., Perl, D. P., Steele, J. C., Chen, K. M., Pierce, J. D., Reyes, P. & Kurland, L. T. (1991). Odor identification deficit of the parkinsonism-dementia complex of Guam: equivalence to that of Alzheimer's and idiopathic Parkinson's disease. *Neurology* **41** (supplement 2), 77–81.

Driscoll, C. T., Baker, J. P., Bisogni, J. J., & Schofield, C. L. (1980). Effect of aluminium speciation on fish in dilute acidifed waters. *Nature* **284**, 161–4.

Ekvall, R. B. (1939). *Cultural Relations on the Kansu-Tibetan Border*. Chicago: The University of Chicago Press.

Elian, M. (1991). Olfactory impairment in motor neuron disease: a pilot study. *Journal of Neurology, Neurosurgery, and Psychiatry* **54**, 927–8.

Figlewicz, D. A., Garruto, R. M., Krizus, A., Yanagihara, R. & Rouleau, G. A. (1994). Absence of mutations in the Cu/Zn superoxide dismutase gene in amyotrophic lateral sclerosis and parkinsonism-dementia of Guam. *Neuro Report* **5**, 557–60.

Flaten, T. P. & Garruto, R. M. (1992). Polynuclear ions in aluminum toxicity. *Journal of Theoretical Biology* **156**, 129–32.

Flaten, T. P., Wakayama, I., Strum, M. J., Kretser, W., Capone, Dudones, T., Bath, D. W., Gallagher, J., Moore, W. & Garruto, R. M. (1993). Natural models of aluminum neurotoxicity in fish from lakes influenced by acid precipitation. In: *Heavy Metals in the Environment–9th International Conference*, vol. 2, ed. R. J. Allan & J. O. Nriagu, pp. 285–8. Edinburgh: CEP Consultants Ltd.

Frisancho, A. R. & Greksa, L. P. (1989). Developmental responses in the acquisition of functional adaptation to high altitude. In: *Human Population Biology: A Transdisciplinary Science*, ed. M. A. Little & J. D. Haas, pp. 203–21. New York: Oxford University Press.

Gajdusek, D. C. (1962). Congenital defects of the central nervous system associated with hyperendemic goiter in a neolithic highland society of Netherlands New Guinea, I: Epidemiology. *Pediatrics* **29**, 345–63.

Gajdusek, D. C. & Garruto, R. M. (1975). The focus of hyperendemic goiter, cretinism, and associated deaf-mutism in Western New Guinea. In: *Biosocial Interrelations in Population Adaptation*, ed. E. Watts, F. E. Johnston & G. W. Lasker, pp. 267–85. The Hague: Mouton Publishers.

Gajdusek, D. C., Garruto, R. M. & Dedecker, R. (1974). Congenital defects of the central nervous system associated hyperendemic goiter in a neolithic highland society of Western New Guinea, V: A note on birth weights and infantile growth rates in the Mulia population. *Human Biology* **46**, 339–44.

Gajdusek, D. C. & Salazar, A. M. (1982). Amyotrophic lateral sclerosis and parkinsonian syndromes in high incidence among the Auyu and Jakai people in West New Guinea. *Neurology* **32**, 107–26.

Garruto, R. M. (1991). Pacific paradigms of environmentally induced neurological disease: Clinical, epidemiological and molecular perspectives. *Neurotoxicology* **12**, 347–78.

Garruto, R. M. & Dutt, J. S. (1983). Lack of a prominant compensatory polycythemia in traditional native Andeans living at 4200 m. *American Journal of Physical Anthropology* **61**, 355–66.

Garruto, R. M., Flaten, T. P. & Wakayama, I. (1993). Natural and experimental models of environmentally induced neurodegeneration: implications for Alzheimer's disease. In: *Alzheimer's Disease: Advances in Clinical and Basic Research*, ed. B. Corain, K. M. Iqbal, M. Nicolini, B. Winblad, H. Wisniewski & P. Zatta, pp. 257–66. London: John Wiley & Sons.

Garruto, R. M., Fukatsu, R., Yanagihara, R., Gajdusek, D. C., Hook, G. & Fiori, C. E. (1984). Imaging of calcium and aluminum in neurofibrillary tangle-bearing neurons in parkinsonism-dementia of Guam. *Proceedings of the National Academy of Sciences* **81**, 1875–9.

Garruto, R. M., Gajdusek, D. C. & ten Brink, J. (1974). Congenital defects of the central nervous system associated with hyperendemic goiter in a neolithic highland society of Western New Guinea, III: Serum and urinary iodine levels in goitrous and adjacent non-goitrous populations. *Human Biology* **46**, 311–29.

Garruto, R. M., Weitz, C. A., Yuan, K. F., Gang, Q., Ma, T. Y., Shen, L. Y., Miao, H. C., Liu, J. C., Lin, L. M. & Chin, C. T. (1991). Study of population isolates at high altitude: the Qinghai project. *American Journal of Physical Anthropology* **12**, 76 (abstract).

Garruto, R. M., Yanagihara, R. & Gajdusek, D. C. (1985). Disappearance of high incidence amyotrophic lateral sclerosis and parkinsonism-dementia of Guam. *Neurology* **35**, 193–8.

Greksa, L. P. (1986). Chest morphology of young Bolivian high altitude residents of European ancestry. *Human Biology* **58**, 427–43.

Greksa, L. P. (1987). Lung function of young Aymara highlanders. *Annals of Human Biology* **14**, 533–42.

Greksa, L. P. (1988). Effect of altitude on the stature, chest depth and forced vital capacity of low-to-high altitude migrant children of European descent. *Human Biology* **60**, 23–32.

Greksa, L. P. (1990). Developmental responses to high-altitude hypoxia in Bolivian children of European ancestry: a test of the developmental adaptation hypothesis. *American Journal of Human Biology* **2**, 603–12.

Greksa, L. P. & Beall, C. M. (1989). Development of chest size and lung function at high altitude. In: *Human Population Biology: A Transdisciplinary Science*, ed. M. A. Little & J. D. Haas, pp. 222–38. New York: Oxford University Press.

Harrison, P. J. (1990). Neurodegeneration and the nose. *Clinical Otolaryngology* **15**, 289–91.

Kidson, C. & Gajdusek, D. C. (1962). Congenital defects of the central nervous system associated with hyperendemic goiter in a neolithic highland society of Netherlands New Guinea. II: Glucose-6-phosphate dehydrogenase in the Mulia population. *Pediatrics* **29**, 364–8.

Kurland, L. T. & Mulder, D. W. (1954). Epidemiologic investigations of amyotrophic lateral sclerosis. 1. Preliminary report on geographic distributions, with special reference to the Marianas Islands, including clinical and pathological observations. *Neurology* **4**, 355–78: 438–48.

Lasker, G. W. (1969). Human biological adaptability: the ecological approach in physical anthropology. *Science* **166**, 1480–6.

Rosen, D. R., Siddique, T., Patterson, D., Figlewicz, D. A., Sapp, P., Hentati, A., Donaldson, D., Goto, J., O'Regan, J. P., Deng, H.-X., Rahmani, Z., Krizus, A., McKenna-Yasek, D., Cayabyab, A., Gaston, S. M., Berger, R., Tanzi, R. E., Halperin, J. J., Herzfeldt, B., Van den Bergh, R., Hung, W.-Y., Bird, T., Deng, G., Mulder, D. W., Smyth C., Laing, N. G., Soriano, E., Pericak-Vance, M. A., Haines, J., Rouleau, G. A., Gusella, J. S., Horvitz, H. R. & Brown, R. H., Jr. (1993). Mutations in cu/zn superoxide dismutase gene are associated with familial amyotrophic lateral sclerosis. *Nature* **362**, 59–62.

Shen, L.-Y., Liu, J.-C., Lin, L.-M., Miao, H.-C., Song, G.-Y., Zan, X.-L., Chin, C.-T., Weitz, C. A. & Garruto, R. M. (1992). Developmental changes in hematological characteristics among Han and Tibetan boys born and raised at high altitude in western China. *American Journal of Human Biology* **4**, 149 (abstract).

Shiraki, H. & Yase, Y. (1975). Amyotrophic lateral sclerosis in Japan. In: *Handbook of Clinical Neurology*, vol. 22, ed. P. J. Vinken & G. W. Bruyn, pp. 353–419. New York: Elsevier Publishing Company.

Strong, M. J. & Garruto, R. M. (1989). Isolation of fetal mouse motor neurons on discontinuous percoll density gradients. *In Vitro Cellular and Developmental Biology* **25**, 939–45.

Strong, M. J. & Garruto, R. M. (1991). Neuron-specific thresholds of aluminum toxicity *in vitro*: A comparative analysis of dissociated fetal rabbit hippocampal and motor neuron-enriched cultures. *Laboratory Investigation* **65**, 243–9.

Strong, M. J., Wolff, A. V., Wakayama, I. & Garruto, R. M. (1991). Aluminum-induced chronic myelopathy in rabbits. *Neurotoxicology* **12**, 9–22.

Van Rhijn, M. (1960). Verspreiding van het endemisch struma in Nederlands Nieuw-Guinea. *Mededelingen van de Dienst van Gezondhidszorg in Nederlands Nieuw-Guinea* **7**, 55.

Weitz, C. A. & Garruto, R. M. (1993). A possible developmental component to chronic mountain sickness. *American Journal of Physical Anthropology* **16**, 206–7 (abstract).

Weitz, C. A. Garruto, R. M., Liu, R.-L., Zhou, J., Shen, L.-Y., Miao, H.-C., Yuan,

K.-F., Gang, Q., Ma, T.-Y., Liu, J.-C., Lin, L.-M. & Chin, C.-T. (1992). Growth of chest size and pulmonary function among Tibetan and Han boys born and raised at 3200 m and 4300 m in western China. *American Journal of Human Biology* **4**, 150 (abstract).

Winslow, R. M., Chapman, K. W. & Monge, C. C. (1990). Ventilation and the control of erythropoiesis in high-altitude natives of Chile and Nepal. *American Journal of Human Biology* **2**, 653–62.

11 *Human biological adaptability with special emphasis on plasticity: history, development and problems for future research*

LAWRENCE M. SCHELL

Summary

The human adaptability paradigm underlies much of the research in human biology conducted over the last thirty years, although explicit application of the paradigm in research has been very rare. Barriers to its use include unstandardized terminology to describe and analyse adaptation, reduced funding for research in adaptability, and the small number of human biologists working in any single area of adaptability. The most serious impediment, however, is lack of agreement among scientists as to how non-genetic adaptation should be measured. Two approaches exist: evaluate physiological and growth variations in terms of their effect on survival and reproduction, as though they are genetic adaptations; or evaluate non-genetic adaptations in terms of their benefit in restoring homeostasis and achieving well-being. The two approaches may be seen as providing two types of proof, with the genetic standard being more difficult to attain.

Special complexities surround the interpretation of plasticity, or adaptation through growth and development. Two models exist to interpret growth: the adaptability model, in which growth is a *means* of adaptation resulting in plastic (ontogenetic) adaptations; and the medical model, in which growth is a *measure* of adaptation. This essay shows how the two models conflict, and notes that the conflict between them cannot be resolved with existing knowledge in human biology. Instead, each model may be limited to a particular area of human biology research. Future research in adaptability may resolve this conflict and develop methods for the replicable measurement of all forms of non-genetic adaptations.

Introduction

In 1969 Gabriel Lasker published an exceptionally articulate review paper. This paper identified three modes of human adaptation; natural selection of genotypes, plasticity (ontogenetic modification), and individual acclimatization. This three-part conceptualization of human adaptability has been very influential on thinking and teaching in biological anthropology. Particularly important was the inclusion of ontogenetic modification as one of the three modes of adaptation. The goal of this chapter is to evaluate this conceptualization of adaptation and variation, taking advantage of twenty-five years of research since Lasker's article, and to examine the impact of Lasker's paper on subsequent research in biological anthropology.

Lasker's article was not the first statement on adaptability (for example, see Harrison, 1966; and for review, Little, 1982). However, it codified in the scientific journal of record in the USA a relatively new framework for interpreting human variability and particularly variation in human growth. This framework had developed over the previous fifty or more years, beginning with Boas' 1911 paper on morphological plasticity and the growth of immigrant children. By mid century, numerous studies comparing migrants and sedentes had dealt with criticisms of the migrant–sedente research design. Their results supported Boas' original study and established the importance of environmental influences on morphology. In 1954 Kaplan summarized research on the plasticity of growth, showing that within genetically homogeneous populations, the growth patterns of sedente and migrant children differed in ways that could best be explained by reference to differences in their environments. The publication of this review in the journal of record for American anthropology indicates that plasticity was now widely accepted in the USA as a *bona fide* phenomenon. It should be noted that Kaplan did not identify plasticity as a means of adaptation.

During the fifty years when migrant–sendente studies established plasticity as a real biological phenomenon, the concept of adaptation to interpret biological variation gained support also (for example, see Coon, Garn & Birdsell, 1950; and Little 1982, for a cogent review of intellectual development at this juncture). In 1964 the International Biological Programme (IBP) proposed studies in human adaptability (Weiner, 1966) and the study of plasticity figured prominently in these. *The Biology of Human Adaptability* (Baker & Weiner, 1966) described '...this world programme in "human adaptability" as one of the main projects of the International Biological Programme' and Harrison (1966) delineated the three modes of adaptation when he described adaptation to high altitude. Thus, the question that developed in biological anthropology from fifty years of research in adaptation and in growth and development was whether flexibility in growth and development had adaptive value. Lasker's 1969 article established the appropriateness of testing the possibility of adaptive growth. It brought together the observations on plasticity with questions of adaptation and presented them in the journal of record for science in the USA.

Measuring the scientific impact of this paper is exceptionally difficult. Standard quantitative approaches include counting citations in articles and in commonly used textbooks in the discipline. As of June 1993, there have been only eight citations of Lasker's 1969 article listed in Science Citation Index. Table 11.1 presents a review of textbooks; although it is not a comprehensive review, it suggests that there has been little explicit use of the article in textbooks as well. Nevertheless, Lasker's article is widely regarded among colleagues as a classic contribution.

It is difficult to explain the small number of research articles that have cited Lasker's 1969 article. Garfield's study of citation patterns in anthropology (1983) concludes that anthropology in general exhibits unusual citation patterns, including long latency before articles are cited. However, he also shows that these characteristics are less common in biological anthropology than in the other subdisciplines of archeology, ethnology and anthropological linguistics. The lack of citation of Lasker's article cannot be explained then as typical of citation patterns in biological anthropology.

It appears, however, that in the surveyed texts (Table 11.1) there is considerable use of the tripartite conception of adaptation, as well as recognition of plasticity as a human characteristic that pertains, in some loosely defined way, to adaptation. Thus, the general view of adaptation and variation may be fundamental to biological anthropology even if Lasker is not cited as its source. In fact, it could be argued that this paradigm in biological anthropology is so fundamental that explicit citation is not necessary.

If the human biological adaptability paradigm can be found in some form in so many introductory textbooks of biological anthropology, why is it not cited in research articles? The disparity between the use of adaptability and plasticity in textbooks and the lack of citation in research articles may indicate that there have been barriers to the application of the human biological adaptability paradigm in the practice of research. By specifying these barriers we hope also to specify the research problems that must be solved for the human adaptability paradigm to move the subdiscipline forward. Furthermore, the disparity brings to the forefront the underlying issue of whether the conceptualization of adaptation that Lasker presented has been useful, and whether it will continue to be so.

Barriers from terminology

One barrier to the application of the human adaptability paradigm is the plethora of terms to describe adaptability, and the use of these terms by different researchers to mean different things. This is confusing and limits

Table 11.1. *Review of textbooks for their treatment of human adaptability*

Author	Explicit model?	Mode I	Mode II	Mode III
Specialized texts on human biology, variation or adaptation				
Baker in Harrison et al. (1988)	Model explicit	Genetic	———— Acclimatization ———— (no distinction explicit)	
Frisancho (1981)	Model explicit	Genetic	Developmental	Acclimatization
Little & Morren (1976)	Model explicit (pp. 6,29) but no citation	Evolutionary	Prolonged	Short-term
Stini (1975)	Model explicit	Genetic	Developmental acclimatization	Acclimatization
Textbooks of physical anthropology				
Relethford (1994)	Model explicit (p. 434) but no citation	Genetic adaptation	Developmental adaptation	Physiological acclimatization
Weiss & Mann (1990)	Model explicit (p. 448)	Genetic	Developmental response	Acclimatization
Poirier et al. (1990)	Model explicit in examples; cites Frisancho (1985)	Adaptation (p. 94)	Developmental acclimatization (p. 548)	Acclimatization (p. 548)
Stein & Rowe (1989)	Explicit model (pp. 151–61) but no citation	Adaptation adjustment	Developmental Adjustment	Acclimatory
Nelson & Jurmain (1988)	Model explicit; cites Lasker (1969)	Biological adaptation	———— Acclimatization ———— (no distinction explicit)	
Johnston (1982)	Model explicit; cites Frisancho (1979)	Genetic adaptation	Developmental adaptation	Physiological acclimatization
Bennett (1979)	Model explicit (follows Mazess, 1975), and a	Genetic adaptation	———— Acclimatization ————	
	2nd model (p. 111)	Genetic homeostasis	Developmental homeostasis	Physiological homeostasis
Damon (1977)	Model not explicit	———— no distinctions explicit ————		
Kelso (1970)	Model not explicit	Adaptation	———— Physiological plasticity ———— (no distinction explicit)	

the exchange of knowledge among independent researchers and the accumulation of knowledge in a field.

The real significance of unstandardized terminology is even greater. Historians and philosophers of science (Scheffler, 1962; Kuhn, 1962) have emphasized the importance of terminology because different terms imply different and usually competing paradigms. When terms develop the same referent in the hands of different researchers, it marks the acceptance of one particular paradigm by those researchers.

Because no paradigm is perfect, all paradigms also include uncertainties, at very least regarding the extent of their application. Thus, a paradigm also identifies a group of problems for researchers to solve. In Kuhn's conception of intellectual change in science this is termed normal science. Thus, standard terminology allows researchers to conceptualize a set of interrelated research problems and their solution solves certain unknowns identified by the paradigm. (To foreshadow a point dealt with much later, an important 'unknown' that Lasker identified in 1969 was whether plasticity can be, or is, adaptive.)

Adaptation

What are the terminological difficulties in human adaptability studies? The definition of adaptation is the most striking. There have been many reviews of the subject and the ground is too great to review again here (see Little, 1982, for one good review and citation of other reviews). Many reviews by anthropologists distinguish between adaptation in a general sense that includes biological and social adjustments to improve wellbeing, and adaptation in a specifically genetic sense that refers to beneficial changes in the frequencies of genes made by populations. Beyond that common distinction there is much disagreement. Frisancho, in his widely used textbook on the subject (1981), is reluctant to define it, stating (p. 2) 'Although the term *adaptation* has been widely used by biologists, social scientists, and laymen, there is not general agreement as to its meaning'.

Although it is true that the term has different meanings stemming from its use in different fields (biology, genetics, psychology, anthropology), each definition of adaptation implies a specific standard for measuring adaptation. Even slight differences in wording can have consequences when the term is operationalized in human biology research. For example, Lasker states 'Adaptation is the change by which organisms surmount the challenges to life' (1969 p. 1481). Taking the definition literally, challenges *to* life are life-threatening, that is potentially lethal if not surmounted. Thus, if an organism is living it has surmounted all past challenges and is, therefore, adapted. Operationalizing a definition with 'challenges *to* life' as the criteria of adaptedness is straightforward: all survivors are equally adapted.

Mazess (1975, p. 10) points out that this is too limiting a definition, '...adaptation, if it is not to be teleological, must mean more than this [mere existence]. The essence of environmental adjustments deemed adaptive seems to be that they are considered RELATIVELY advantageous, beneficial, or meritorious, or that they are to a degree necessary' (capitalization in original). By rewording Lasker's definition slightly to challenges *of* life, its operationalization now involves assessing which putative adaptations are, in fact, relatively beneficial. (How to measure 'relatively beneficial' is described below, in a section on measurement problems.) The key point in this discussion of Lasker's definition and its rewording is that the definition of a key concept has significant ramifications for the conduct of research. Differences in the definition of adaptation mean there are different ways to measure adaptation (e.g. relative benefit vs. survival), different short-term research goals, and even different paradigms.

To whom or to what does adaptation occur: the level problem

A second area of basic terminological disagreement concerns where adaptation occurs, or to whom: the organ system, organism, the population, or the species? At what level is adaptation occurring; consequently, at what level should it be measured?

The answer is important first because different levels of adaptation imply different means of assessing the relative benefit of an adaptive process or form. Thus, in research on adaptability, the level on which adaptation is being studied must be specified and a level-specific measure of relative benefit employed.

The question of levels also begs another more complex one, the relationship between responses at different levels. Even when adaptation at one level is assessed with a level-specific measure of relative benefit, adaptation at one level may have consequences at other levels and these will need to be considered also. For example, slow physical growth and small body size for age may represent a beneficial adjustment for the population with insufficient nutrients, but for the individual, small size and past slow growth are usually associated with poorer health in the past and future (Tanner, 1966, 1979, 1986). Thus, the human adaptability paradigm specifies at least two research problems: (1) what is the relationship between adaptation at one level and that at another, an issue researched empirically with regard to the relationship between body size on the one hand and survival or reproduction on the other (see Clark & Spuhler, 1959; Martorell *et al.*, 1981; Mueller, 1979; Shapiro *et al.*, 1983; Waaler, 1984); and (2) is one level of adaptation (e.g. the population level in the growth example given) paramount over levels of less organizational complexity?

What are the forms of adaptation?

Misunderstandings among researchers also arise through unstandardized terminology to describe the forms of adaptation. Adaptation has been described in terms of 'modes' or 'types' or 'levels' and sometimes the terms are used interchangeably. In 1969 Lasker uses *modes* to refer to adaptations arising through different mechanisms, as in, 'The three modes of adaption (selection, plasticity, and acclimatization)...' (p. 1485); earlier (p. 1481) he uses *levels* similarly, 'Adaptation occurs at three levels: (i) selection of the genotype, (ii) ontogenetic modification, and (iii) physiology and behavioral response'. Whatever terms are used, Lasker and other researchers consistently distinguish two parameters of adaptation: (1) the modes of biological adaptation (selection, plasticity, acclimatization) and (2) the level of biological aggregation on which adaptation occurs. For the remainder of this chapter the term *mode* is used to refer to adaptations that arise through different mechanisms (genetic, growth and development, physiological, or behavioral mechanisms), and the term *level* to refer to the biological aggregation at which adaptation is evaluated. Thus, levels are species, populations, individuals, organ systems, etc. (see Mazess, 1975, for elucidation of the levels of organization complexity).

Distinguishing between modes and levels is important. For example, Lasker (1969) draws a connection between level, in the sense of individual, population and species, and mode (genetic, ontogenetic and physiological/behavioral) of adaptation (p. 1481): 'As one goes from interspecific differences to individual differences within the human species, the chief emphasis shifts from the first level [(i) selection of the genotype] to the second [... (ii) ontogenetic modification ...] and third [... (iii) physiology and behavioral response ...] ' (bracketed expressions added). One benefit of consistent terminology that distinguishes levels from modes is that it permits a test of Lasker's hypothesis of a connection between the two dimensions, and consistent terminology in general will facilitate formulating specific problems for research with the human adaptability paradigm and comparisons of results across studies.

Having distinguished levels from modes, it is clear from a review of several texts and review articles dealing with adaptability (see Table 11.1) that the terms used to refer to the modes are not standardized. There are several consistencies in usage, however, which suggest that standard uses are developing. Virtually all writers identify a genetic mode, referring to adaptation through changes in gene frequencies by the mechanism of natural selection. Short-term reversible responses are termed acclimatization or physiological adaptations.

Intermediate between genetic and acclimatization modes is the third and, according to Lasker, the most recently identified mode which he termed plasticity, and elsewhere, ontogenetic modification. It has a variety of rough synonyms in the various review articles and texts in which the concept is elucidated, including developmental acclimatization and long-term acclimatization. Ontogenetic modification and its synonyms have slightly different connotations but all (1) refer to the adaptations themselves, and not the capacity to adapt, that are (2) made through the mechanisms of growth and development, which are (3) irreversible but (4) not heritable. For the remainder of this chapter we shall use Lasker's original term, ontogenetic modification, and leave the elaboration of this mode's conceptual affinities with other modes as a matter for specific empirical research.

It should be noted that unstandardized terminology is typical of fairly young paradigms and thus physical anthropologists may take the plethora of terms not as a fatal flaw in the paradigm itself but as an indicator of the intellectual maturity of the paradigm. However, an important consequence of unstandardized terminology has been that the measurement of adaptation itself has been inhibited.

Problems of measurement

In science, it is crucial that different researchers are able to make accurate and replicable measurements of the key variables being studied. Thus, the creation of measurement tools for assessing adaptation is a crucial task for all researchers using the adaptability paradigm. It should be clear, judging from the discussion of adaptation's definition and the plethora of terms to define the parameters of its operation (e.g. modes and levels), that exact tools to measure adaptation have not yet been developed.

The absence of tools for exact measurement is unfortunate but not fatal. The development of measurement tools can be viewed as a stage in the process of scientific development. At this point, even approximate measurement tools are better than none since measurement scales can improve with empirical testing. The qualitative scales employed currently to assess adaptation can be considered important prototypes for more precise and replicable scales to be developed.

The measurement of adaptation must follow closely from its definition. Here we accept Lasker's definition but as modified earlier: adaptation is the change(s) an organism makes to surmount the challenges of life. This definition means that existence alone is an insensitive measure of adaptation, there being variation in how well life's challenges are met, and that the measurement of this response is the measure of adaptation.

Genetic adaptation

Measuring genetic adaptation is fairly straightforward. Representation in the next generation can be used as a scale to judge the suitability and success of genetic adaptations. Differences in representation can be attributed to prereproductive mortality and/or differential fertility, and methods exist to express these entities in quantitative terms (Harrison, 1988).

Despite the familiarity of these measurement methods it should be noted that there are very few documented cases of genetic adaptation in humans. Perhaps such adaptation rarely occurs, or perhaps it is very difficult to detect. Detection may require assessing prereproductive mortality at specific time periods, as for sickle cell status and malaria (Allison, 1964), or the existence of detailed records that permit observations with considerable time depth.

Non-genetic adaptation

When the definition of adaptation is enlarged to include all three modes of adaptation, a fundamental problem arises: what is the appropriate scale of measurement? The genetic standard, i.e. representation in the next generation, may not be appropriate for measuring the adaptive value (relative benefit) of acclimatizations. Some physiologic responses that are commonly termed adaptations, such as pupillary dilation in response to light, would not meet the standard since it is impossible to know whether the response actually increased representation in the next generation. Responses such as these are deemed adaptations because they improve or restore functioning in the face of a stressor, but it should be noted that usually they do so in theory only. Rarely are there demonstrations of improved functioning revealed through actual testing. Nevertheless, such responses are commonly termed adaptations. Thus, 'representation in the next generation', although it is an appropriate measurement scale for the genetic mode of adaptation, is not necessarily appropriate for non-genetic modes.

Acclimatization

Many researchers have adopted the general, tripartite definition of adaptation developed here, or at least a bipartite one that includes both genetic and non-genetic modes. Such definitions imply that measuring scales for adaptation can be broadly defined. In an important review of the subject, Little (1982, p. 406) states 'It would appear that adaptation as a state in the individual can be measured according to both (a) survival and general well-being and (b) reproductive success'. Use of survival as a measure of adaptation has been criticized by Mazess (1975) as too limiting (discussed earlier) as it does not allow for the measurement of *relative* benefit; in other words, it is not scalable. This leaves 'wellbeing' as the only alternative to the genetic standard. Measuring the amount of 'wellbeing' that results from a non-genetic adaptation presents difficulties also.

Frisancho (1981) deals extensively with adaptations in his textbook on the subject and has adopted two measurement scales: the customary one for genetic adaptation, and another for ontogenetic modification and acclimatization, the modes that comprise functional adaptation. According to Frisancho, functional adaptations can be assessed in terms of their success in restoring homeostasis and reducing the strain imposed by a stressor (1981, p. 5). This is an appropriate scale of measurement considering the text's emphasis on physiological adjustments to the environment. As long as the focus is upon assessing the adaptive value of physiologic responses, there is a clear intellectual connection to the work of Prosser (1964), Folk (1974) and the discipline of environmental physiology generally, where homeostasis is a central concept.

If, on the other hand, the genetic standard is employed for assessing acclimatization, there are certain consequences for the study of adaptability. First, it becomes necessary to test each response for its value in survival and/or reproduction. This entails a life-span perspective and a longitudinal research design where individuals are followed to collect data from several points in the life-span. Also, we are severely restricted as to what we can call an adaptation. Using the genetic standard, we must refrain from calling a particular 'response' an 'adaptation' unless its value for reproduction or survival has been demonstrated. Thus, such responses as the change the eye makes to changes in light cannot be termed 'adaptations', but only 'responses'. (Some responses we observe may have been adaptations in the past by permitting survival or greater fertility then, but these advantages are untestable now and their status as 'adaptations' becomes questionable with the genetic standard.)

To conclude, applying the genetic standard to non-genetic adaptation imposes limitations that seem highly restrictive and demanding. Furthermore, the genetic standard is not commonly used to assess the relative benefit of acclimatizations. Once the non-universality of the genetic standard is recognized, its application to measure the relative benefit of all non-genetic adaptations becomes questionable.

Plasticity, ontogenetic modification and wellbeing

At present there are no methods available to assess the relative benefit of ontogenetic modifications that are as widely accepted as are the methods developed for assessing acclimatizations. The development of scales for measuring the adaptive value of ontogenetic modifications has been hindered by uncertainty in interpreting growth patterns generally.

Two interpretations of growth and development Variation in growth and development due to environmental factors has been subject to two very different interpretations. One is the interpretation provided through the human adaptability paradigm where ontogenetic modification is a means of adaptation. The second interpretation is based on the medical model employed in public health, pediatrics, toxicology, and nutritional science. From this viewpoint growth reflects the health of the organism: growth to the limit of the organism's genetic potential signifies health, whereas retarded growth signifies ill-health, toxic insult or nutritional stress. In public health, birth weight data are often employed to indicate the general healthfulness of the population and the success of health services. In pediatrics, physical growth along the lines of the community standard is the common expectation and in toxicological studies poor growth or weight loss are fundamental signs of toxicity. Thus, when the medical model is employed, growth is a measure of health.

The medical model can be applied to measure adaptation. Rene Dubos linked health to adaptation in his book (1965, p. xvii), '... states of health or disease are the expressions of the success or failure experienced by the organism in its efforts to respond adaptively to environmental challenges'. This perspective has been adopted by biomedical anthropology (McElroy & Townsend, 1979). It is also entirely consistent with the use of growth and development to measure 'wellbeing', a commonly cited alternative or a supplement to the genetic standard for assessing adaptation (see Mazess, 1975; Little, 1982). In sum, the medical model places growth as a *measure* of adaptation because growth reflects the health of the organism, and in aggregation, of the community also.

Using this model, individuals could be assessed as to how well they have adapted by how well they have grown *given their genetic potential for growth*, and populations could be compared *if they have similar genetic potentials for growth*. Indeed, it may be easier to compare and assess populations than individuals. Individuals vary considerably in growth potential and this makes it difficult to determine when an individual's growth potential is not met. There is less variation in growth potential among populations and this allows us to make reasonable expectations of average growth patterns. In short, the medical model of growth interpretation permits the use of growth as a measure of health *and* of adaptive success.

The human adaptability paradigm, on the other hand, views growth as the mechanism of ontogenetic modification (plasticity). Growth patterns can be a mode of adaptation, a way for the organism to surmount the challenges of life. In this context growth is a *means* of achieving an adapted state rather than a result of that adaptation.

There are only a few examples of probable ontogenetic adaptation. The study of morphology in relation to thermal extremes is one example. Body size and shape are regarded as playing an important role because they pertain to the ratio of surface area to body mass (Baker, 1988; Frisancho, 1981). Modifications arising through growth and development that change the ratio of surface area to body mass to suit the temperature extreme in a given environment may be regarded as adaptations. In hot tropical environments, growth patterns that reduce size and or increase linearity may be regarded as adaptive. Eveleth (1966) observed that children of European descent living in Rio de Janeiro were lighter, leaner and had greater linearity in comparison with children of European descent living in temperate climates of North America. Since these two modifications should reduce heat stress, they may be considered adaptive.

It is noteworthy that in this example there is no actual measurement of relative benefit that growth provides as an adaptation to heat stress. The conclusion of adaptation rests on the theory that reducing size and increasing linearity reduces heat stress and contributes to homeostasis. This assessment of adaptation is similar to the method of assessing acclimatizations. The ontogenetic modification in question reduced, at least in theory (i.e. without its actual measurement), physiological strain and facilitated homeostasis.

Adaptation to high-altitude hypoxia is another test case of ontogenetic adaptation because many morphological and functional characteristics that are distinctive of high-altitude native adults develop during growth. (Baker, 1969; Beall, 1981; Clegg, et al. 1972; Clegg, 1978; Frisancho, 1969; Frisancho & Baker, 1970; Greksa, 1990; Haas, 1976, 1980; Harrison, 1966; Moore, 1990). Frisancho (1976) has proposed an hypothesis, now termed the developmental adaptation hypothesis (Greksa 1990), to explain the unique morphological and functional growth pattern as well as the adult phenotype of high-altitude Andean natives. He proposed that the pattern of accelerated growth of the oxygen transport systems and the slowed growth of body size is a developmental response that results in functional adaptation to high altitude hypoxia.

Recently Greksa (1991) reviewed the evidence for the development of physiological adaptations to high-altitude hypoxia and found considerable evidence to support its presence generally. However, he also found indications of population differences in developmental adaptations. For example, among Quechua Indians of Peru, FVC (forced vital capacity, a measure of lung volume) was similar in high-altitude natives and in males who had migrated to high altitude as children, and their FVC was greater than that of adult males who had migrated as adults (Frisancho, Velasquez

& Sanchez, 1973). Likewise, among Aymara Indians of Chile and Bolivia, FVC increases significantly with altitude (Mueller *et al.*, 1978), but perhaps not as greatly as among the Quechua (Greksa, 1991). Among Tibetans, FVC was not significantly greater among highland residents compared with sea-level ones (Beall 1984; Beall, Strohl & Brittenham, 1983). Greksa (1991) observed that Europeans born and raised at high altitude in Bolivia have smaller FVCs than Andean or Himalayan highland natives. However, he also observed that the Europeans who had migrated to altitude had smaller FVCs than those who had been born and raised there, suggesting significant developmental adaptation in the latter group. Thus, the development of higher FVC may represent a form of ontogenetic adaptation to altitude, but it is an adaptation that some populations make extensively and others perhaps very little.

In summary, the examples from studies of developmental responses to thermal stress and to high-altitude hypoxia are sufficient to indicate that plasticity must be recognized as a possible mode of adaptation. Thus, the adaptability model of growth cannot be dismissed in favor of the medical model.

Conflicts between the medical and adaptibility models   The two interpretations of growth, following from the medical and the adaptability models of plasticity, conflict. Growth cannot be simultaneously a means to adaptation *and* a measure of adaptation, for how would growth as a means of adaptation be assessed, with growth as the measure? If growth were the measure of adaptation, small size for age and slow maturation would never be classed as adaptive: such growth falls short of the organism's growth potential, according to the medical model.

Although the two interpretations appear contradictory when juxtaposed, each interpretive framework is compelling when considered alone, and there are solid examples of growth as adaptation and of growth as a measure of health. Given the contradiction in the two interpretations of plasticity, difficulty in applying the human adaptability paradigm is understandable. Some resolution of the conflict in interpretations may be obtained by recognizing commonalities in the contexts in which each of the two interpretive frameworks have been applied.

One commonality concerns the stressors themselves. Interpretations of growth as a mode of adaptation cluster in studies of adaptations to the physical environment (for review see Beall, 1982). Numerous studies of populations in Guatemala (Johnston *et al.*, 1976), Brazil (Eveleth, 1966), and the U.S. (Newman, 1953; Newman & Munro, 1955) as well as global surveys (Roberts, 1953) have interpreted growth patterns as part of a morphological adjustment to temperature in conformity to Bergman's and Allen's rules.

Examples of the use of growth data as an indicator of health are in evaluating the impact on human biology of different dietary regimes (Himes, 1991), pollution and toxic insults (Schell, 1992; Schell & Ando, 1992), socioeconomic status (Bielicki, 1986) and disease (many, but not all, infectious diseases produce deficits in growth; proper therapy is expected to restore good growth). Thus, the use of growth as an indicator of health is common when the individual is challenged by features created by or substantially modified by the actions of humans. As a counterexample, researchers of growth and adaptation would not suggest that the poor growth of slum children represents a *successful* adaptation to slum conditions, but rather that it is an unfortunate consequence of those conditions.

There is another commonality to the way in which the two interpretations of growth patterns have been applied. Studies of growth as adaptation to stressors of the physical environment have focused on one stressor, whether it is temperature, hypoxia or some other stressor. Studies that interpret growth as a response and a measure of health generally view growth as a summary of the organism's ability to cope with many features. This may be due to the fact that the human-made environment is not as easily broken down into component stressors as the physical environment, or there is less agreement on how to do it.

In short, when judging the impact on well-being or health of different diets, exposure to a pollutant, or even family size and birth order, 'good' growth (e.g. growth within the limits of an individual's genetic potential) is commonly taken as an indicator of success. On the other hand, when judging the impact on biology of features of the physical environment, the growth trajectory becomes a biological resource that can flex, turning slow or fast, as might be required to improve survival and fitness.

The creation of a rule to guide the applications of the medical model and the adaptability model in interpreting growth patterns may reconcile the two interpretations when the stressor being considered is singular (either human-made or a feature of the natural environment): apply the medical model to studies of adaptation to a feature of the human-made environment, and the adaptability model to studies of adaptation to a feature of the physical environment.

This approach is consistent with the idea of a hierarchy of response by the three modes of adaptation (Thomas, 1975; for earlier formulation of this idea see Slobodkin, 1968, and Bateson, 1963). Behavioral and physiological adjustment constitute the most immediate avenues of response to a stressor, but if this is insufficient, a response through ontogenetic modification may occur. If this is inadequate, limitations on individuals' survival or fertility may alter the genetic composition of the populations and make more common those genes that permit survival and reproduction. In this view, ontogenetic modifications represent an adjustment to a stressor after behavioral and acclimatization modes have failed to completely buffer a stressor. The medical model becomes appropriate as features of the physical environment become less stressful through the action of the behavioral mode of adaptation (i.e. culture and technology) and growth patterns come to summarize the success of these forms of adaptation. In some cases, the social–behavioral–technological adjustment produces a new stressor, e.g. pollution (Schell, 1992), infectious disease (Armelagos & Dewey, 1970), unequal distribution of macro- or micro-nutrients and nutritional stress (Greene, 1974, 1977); in these cases the medical model is commonly applied.

The model developed for the single-stressor analysis may be limited because in most situations the environment includes multiple stressors to which human beings' growth responds *in toto*. Even such an extreme environment as the Peruvian altiplano is characterized by the stressors of cold, aridity, wind, and a suboptimal diet, as well as hypoxia. In this case adaptation to some stressors may entail different patterns of growth. For example, adaptation to cold may entail larger body size while adaptation to hypoxia may entail slower growth and smaller body size for age. Thus, in the multistress environment growth is more likely to be seen as a summary measure of health rather than as a means for the individual to adapt to each of the stressors.

The use of growth as a summary measure of health in a multistress environment has a precedent. It is similar to the use of survival and reproduction as a measure of genetic adaptation. In the latter, differences in representation in the next generation from either differential prereproductive mortality or differential fertility may be due to a combination of an individual's genetic features, some contributing positively and others negatively. Representation is a summary statement of the organism's reproductive performance and when aggregated, of the population's performance also. Growth, when interpreted with the medical model, also summarizes the combined contributions of the organism's positive and negative responses rather than disentangles them. In this way it is similar to the genetic standard for assessing adaptation.

Empirical verification

Empirical verification of adaptation through ontogenetic modification has been difficult. Unresolved issues of terminology and measurement make it unclear just what empirical verification entails; the level at which adaptation should be measured (the individual or the population) is often not specified, and plasticity is open to at least two interpretations. If we use the genetic standard, a study of the plasticity mode of adaptation requires more than the description of growth but also measuring the essentially irreversible adult phenotype, and the influence on survival and reproduction, with this last step being the difference between describing a growth *response* to an environmental stress and identifying plastic *adaptation* to an environmental stress.

Using the genetic standard, several issues are present in investigating the plasticity mode of adaptation. One is how to identify the presence or absence of a response from exposure to a stress as a subadult. A second is how to measure the appropriate aspect of the adult phenotype, determine its reversibility and distinguish it from the ageing process. The third is how to measure the influence on that aspect of survival and reproduction which occurs decades after the response is expressed. With the genetic standard, an adequate test of the hypothesis that a particular characteristic of subadults represents the plasticity model of adaptation requires adopting a lifespan theoretical perspective, rather than one focused on child growth and development. A lifespan perspective evaluates the influence of characteristics at one stage of the life cycle upon subsequent stages.

Several studies have sought to determine the adaptive value of a growth pattern by examining the relationship between size for age on the one hand, and mortality, morbidity or fertility on the other (Clark & Spuhler, 1959; Damon & Thomas, 1967; DeScrilli *et al.*, 1983; Martorell *et al.*, 1981; Mueller, 1979; Shapiro *et al.*, 1983; Waaler, 1984). An important feature of these studies is that they focused on the individual level in adaptation by measuring both the growth (size) and the outcome (survival, fertility, etc.) in individuals. They did not consider the benefit at the population level from an adaptation made at the individual level.

Such studies have produced a variety of assessments, including linear and curvilinear relationships between size and mortality. The variety of relationships does not necessarily indicate confusion, faulty research designs, etc., but instead that what is adaptive in one context, at least by the genetic standard, is not necessarily adaptive in another. For example, a measure of body mass, such as weight for height, may have a different impact on fertility in well-nourished populations than in malnourished ones. In malnourished populations, greater than average weight for height signifies more energy reserves and these can be helpful in prereproductive survival and in reproduction; in a well-nourished population, greater than average weight for height may reach obesity with all the pregnancy risks that obesity entails.

Given the complexities in interpreting the adaptive value of biological responses, the issue of ontogenetic adaptation is very poorly summarized by the question 'Is bigger really better?' Although pithy, this question does not take into account any of the issues involved in measuring ontogenetic adaptations. It does not consider the context of adaptation or the level on which it occurs and is measured. It does serve, however, to test researchers' ability to define adaptation and measure it.

One of the most problematic tests of plastic adaptation is the issue of nutritional adaptation and the 'small but healthy' hypothesis. In 1969 Lasker included nutritional adaptation as one of the examples of human adaptability. In 1982, Seckler posed a 'small but healthy' hypothesis also. This hypothesis reasons that populations with short stature are adapted to undernutrition because the individuals' small body size allows them to subsist on fewer calories, but without a loss of function. This is a highly controversial interpretation of small body size (see Bogin in this volume for additional treatment of this issue; also Stinson, 1992, for review).

Although 'small but healthy' is an intuitively appealing hypothesis and one that is free of a Western value system that seems to value size in general, there are several problems inherent in the statement of the hypothesis that raise important questions about empirical testing. The benefit derived from subsisting on fewer calories could be measured at the individual or the population level. If the latter, are there benefits for the population from changes in growth (size) that occur at the level of the individual? None of the examples of empirical testing with the genetic standard just cited above has examined this possibility, and there are no standard methods for assessing such cross-level hypotheses. Even if benefits for the population could be demonstrated, how would these be weighed against effects on the individual level?

If the hypothesis is restated to imply that the adaptive benefits of small size affect the individual, the hypothesis is more testable. If the genetic standard were not required, and wellbeing were substituted as the measure of adaptation, several areas of function could be investigated simply by comparing functional ability among children with different growth patterns, usually size for age (again see Bogin, this volume). Scrimshaw & Young (1989) show that there is a cost in functional terms to the reduced dietary intakes that result in smaller body size during growth and ultimately smaller adult size as well. They conclude that smaller body size is an accommodation to reduced intakes, rather than an adaptation, because the reduction permits or prolongs survival, but it also has significant adverse biological consequences for the individual.

Another study examining the relationship of nutrition to size and functioning is being conducted by R. Martorell & J. D. Haas in Guatemala. Theirs is a longitudinal study of over 1600 participants from an earlier study of nutritional intervention during the preschool years. They hope to determine whether the earlier supplementation benefited later function, such as work capacity or cognitive development. To add the genetic standard to this study would require data on mortality and fertility on the same cohort who received supplementation as children.

Summary
There is no single interpretation of growth, but two competing interpretative frameworks. Both appear appropriate but in quite different contexts. Examples of growth as an adaptive dimension exist in studies of adaptation to features of the physical environment and to singular features in particular. In response to human made features and in multistress environments, the medical model of growth interpretation is more common. Proof of the adaptive value of an ontogenetic modification occurs at two levels: a first level where the feature is judged adaptive because it facilitates, in theory or by direct measurement, homeostasis or improved functioning; and a second level where the feature is shown to promote survival and fertility. Few examples of ontogenetic modification reach the second level of certainty because empirical verification of the adaptive value of a growth pattern is difficult. It requires a lifespan perspective with considerable follow-up to determine the value of a growth feature on prereproductive mortality and on fertility.

Funding and progress with the human adaptability paradigm
Funding plays a crucial role in scientific research. The direction that research takes depends on what funds are available, and new areas of research can open or close as funding for the areas change. Inadequate funding has impeded research on human adaptability.

A significant impetus for Lasker's (1969) paper was the International Biological Programme (IBP) and its component, the Human Adaptability Project. The explicit goal of Lasker's paper was to explain the ecological approach in biological anthropology that was the central to the Human Adaptability Project. Through the IBP, numerous studies in human adaptability were conducted between 1967 and 1972 (Collins & Weiner, 1977; Roberts, 1993; Little, 1982). Funding for several of the IBP human adaptability research projects conducted by US scientists came from the National Science Foundation (NSF) and the National Institutes of Health (Baker & Eveleth, 1982), and for British-led research projects in Ethiopia and New Guinea, from the Royal Society of London.

After the IBP terminated in 1974, continued work in human adaptability depended on continued funding for biological anthropology. At this crucial juncture in the mid-1970s, funding in the USA for biological anthropology leveled off after growing from the 1950s to the mid-1960s (Baker & Eveleth, 1982). At the same time the importance of human population biology research waned and an increasing percentage of the NSF budget went for research in primatology (Baker & Eveleth, 1982). In short, since the early 1970s support for human adaptability research has decreased.

Roberts (1993) has recently reviewed the state of international funding for research in human biology. Good prospects for current research exist in several areas of interest to human biologists, including the human genome, tropical populations, and the impact of global change. Cross-cutting these tropical areas is an emphasis on ecology and interdisciplinary approaches, and a focus on health.

Support for research to explicitly test anthropological theory is unlikely to receive support from many funding agencies today. Indeed, the history of funding in biological anthropology shows that, excepting university support for strictly anthropological projects, biological anthropology research has usually been tied to goals outside of biological anthropology.

In the USA the only sources devoted entirely to anthropology and likely to appreciate and fund projects of importance to bioanthropological theory are the National Science Foundation and the Wenner–Gren Foundation. Their budgets may be smaller than they once were after considering inflation, the growth of new specialities (e.g. primatology), and the number of anthropologists in the field.

In short, the lack of progress in testing and refining the human adaptability paradigm may be due to a lack of funding and sustained, focused research during the past twenty-five years rather than to inherent deficiencies in the paradigm itself. Empirical studies to develop rigorous measurement scales of adaptation, or to test relationships between adaptive success at different levels, to cite but two examples, cannot be conducted without substantial funding. It is not surprising therefore that progress in the traditional areas of human adaptability research has been slow since the 1960s. At the same time, funding for biomedical projects using the medical model of growth interpretation is substantial and more research is being done with this interpretative framework.

The scope of bioanthropology
Another possible impediment to substantial progress with the human adaptability paradigm is the small number of researchers working on any specific research problem. Human biologists study an incredibly wide

range of subjects. The number working on any one area of human adaptability is quite small. More progress may be possible with more researchers devoted to one problem.

The research area within human adaptibility that arguably has made the most progress since 1969 is high-altitude studies. Coincidentally, this also is an area in which a great many researchers have worked, perhaps more so than in any other single area of human adaptability. This may be due to the organization of particular graduate programs and the explicit delineation of research problems in that area, two interrelated forces.

Conclusions

The human adaptability paradigm developed through research on migration, plasticity, race and adaptation. Its articulation in several forms by several authors in the 1960s established it as a fundamental paradigm in biological anthropology, along with evolution. However, research in the area did not develop once funding for anthropology and academic human biology declined. Crucial issues of terminology and measurement scales had not been resolved when funding declined. Following the IBP-sponsored Human Adaptability Program research, smaller projects, funded by individuals, universities, or as part of health-related research, did not have a common, standard set of methods or goals. Progress in any single area of adaptability studies was slow relative to progress in the laboratory sciences.

Among the issues unresolved at the close of the IBP was uncertainty as to the interpretation of ontogenetic modification as a mode of adaptation. Although responses through growth and development may include adaptive patterns, there is no agreement as to how they may be distinguished from non-adaptive patterns. This follows from a lack of agreement on the framework for interpreting plasticity and a scale by which to measure ontogenetic adaptations.

The two interpretations of growth patterns, the adaptability and the medical models, appear contradictory and competing. However, consideration of the contexts in which they have been applied successfully indicates that they are rarely used in identical situations. As rules to guide the application of each growth interpretation are developed, a standard for assessing ontogenetic modifications in different contexts may be developed also.

In human biology, as in other sciences, the development of tools that provide a replicable scale of measurement to operationalize the key concepts of the discipline in the conduct of empirical research is essential. Expressions of the human adaptability paradigm by Lasker, Harrison and others in the 1960s and early 1970s codified the three means of biological adaptation and directed our attention to issues in their measurement and interrelationships. However, these issues have remained. Their resolution has been hindered by unstandardized terminology, reduced funding for research adaptability, and the relatively small number of scientists working on any single specific problem in adaptability. In short, lack of progress in human adaptability research may not be the fault of the paradigm itself, but may relate to our ability to apply it by answering the questions it has posed.

Acknowledgements
This chapter was written in response to a paper delivered by Dr. Cynthia Beall at the 62nd annual meeting of the American Association of Physical Anthropologists, Toronto, Canada. The author thanks Drs M. Little, R. Huss-Ashmore and S. Ulijaszek for comments on an earlier version of this manuscript, and Dr. G. W. Lasker for his inspiration generally and his many helpful suggestions over the years.

References
Allison, A. (1964). Polymorphism and natural selection in human populations. *Cold Spring Harbor Symposia on Quantitative Biology* **29**, 137–49.
Armelagos, G. & Dewey, X. (1970). Evolutionary response to human infectious disease. *Bioscience* **20**, 271–5.
Baker, P. T. (1969). Human adaptation to high altitude. *Science* **163**, 1149–56.
Baker, P. T. (1988). Human adaptability. In: *Human Biology* 3rd edn, ed. G. A. Harrison, J. M. Tanner, D. R. Pilbeam & P. T. Baker, pp. 439–543. Oxford University Press.
Baker, T. S. & Eveleth, P. B. (1982). The effects of funding patterns on the development of physical anthropology. In: *A History of American Physical Anthropology, 1930–1980*, ed. F. Spencer, pp. 31–48. New York: Academic Press.
Baker, P. T. & Weiner J. S. (eds) (1966). *The Biology of Human Adaptability*. Oxford: Clarendon Press.
Bateson, G. (1963). The role of somatic change in evolution. *Evolution* **17**, 529–39.
Beall, C. J. (1981). Optimal birthweights in Peruvian populations at high and low altitudes. *American Journal of Physical Anthropology* **56**, 209–16.
Beall, C. M. (1982). An historical perspective on studies of human growth and development in extreme environments. In: *A History of American Physical Anthropology, 1930–1980*, ed. F. Spencer, pp. 447–65. New York: Academic Press.
Beall, C. M. (1984). Aging and growth at high altitudes in the Himalayas. In: *The People of South Asia*, ed. J. R. Lukas, pp. 365–85. New York: Plenum Press.
Beall, C. M., Strohl, K. P. & Brittenham, G. M. (1983). Reappraisal of Andean high altitude erythrocytosis from a Himalayan perspective. *Seminars in Respiratory Medicine* **5**, 195–201.
Bennett, K. A. (1979). *Fundamentals of Biological Anthropology*. Dubuque, Iowa: Wm. C. Brown.
Bielicki, T. (1986). Physical growth as a measure of the economic well-being of

populations: The twentieth century. In: *Human Growth* (2nd edn), vol. 3, *Methodology, Ecologic, Genetic, and Nutritional Effects on Growth*, ed. F. Falkner & J. M. Tanner, pp. 283–307. New York: Plenum Press.

Clark, P. J. & Spuhler, J. N. (1959). Differential fertility in relation to body dimensions. *Human Biology* **31**, 121–37.

Clegg, E. J. (1978). Fertility and early growth. In: *The Biology of High Altitude Peoples*, ed. P. T. Baker, pp. 65–116. Cambridge University Press.

Clegg, E. J., Pawson, G., Ashton, E. H., & Flinn, R. M. (1972). The growth of children at different altitudes in Ethiopia. *Philosophical Transactions of the Royal Society of London* **B264**, 403–37.

Collins, K. J. & Weiner, J. S. (eds) (1977). *Human Adaptability. A History and Compendium of Research in the International Biological Programme*. London: Taylor & Francis.

Coon, C., Garn, S. M. & Birdsell, J. B. (1950). *Races: a Study of the Problems of Race Formation in Man*. Springfield, Illinois: Thomas.

Damon, A. (1977). *Human Biology and Ecology*. New York: W. W. Norton.

Damon, A. & Thomas, R. Brooke (1967). Fertility and physique–height, weight, and ponderal index. *Human Biology* **39**, 5–13.

DeScrilli, A., Bossi, A., Marubini, E. & Caccamo, M. L. (1983). Neonatal morbidity risk: a study of the interrelationship to birthweight and gestational age in four Italian centres. *Annals of Human Biology* **10**, 235–46.

Dubos, R. (1965) *Man Adapting*. New Haven, Connecticut: Yale University Press.

Eveleth, P. B. (1966). The effect of climate on growth. *Annals of the New York Academy of Science* **134**, 750–9.

Folk, G. E. Jr. (1974). *Textbook of Environmental Physiology*. Philadelphia: Lea & Febiger.

Frisancho, A. R. (1969). Human growth and pulmonary function of a high altitude Peruvian Quechua population. *Human Biology* **41**, 365–79.

Frisancho, A. R. (1976). Growth and morphology at high altitude. In: *Man in the Andes: a multidisciplinary study of high-altitude Quechua natives*, ed. P. T. Baker & M. A. Little, pp. 180–207. Stoudsburg: Dowden, Hutchinson & Ross.

Frisancho, A. R. (1981). *Human Adaptation. A Functional Interpretation*. Ann Arbor, Michigan: University of Michigan Press.

Frisancho, A. R. & Baker, P. T. (1970). Altitude and growth: a study of the patterns of physical growth of a high altitude Peruvian Quechua population. *American Journal of Physical Anthropology* **32**, 279–92.

Frisancho, A. R., Velasquez, T. & Sanchez, J. (1973). Influence of developmental adaptation on lung function at high altitude. *Human Biology* **45**, 583–94.

Garfield, E. (1983). Current Comments. *Current Contents* **37**, 5–12.

Greene, L. (1974). Physical growth and development, neurological maturation and behavioral functioning in two Ecuadorian communities in which goiter is endemic. II. PTC taste sensitivity and neurological maturation. *American Journal of Physical Anthropology* **41**, 139–52.

Greene, L. (1977). Hyperendemic goiter, cretinism, and social organization in highland Ecuador. In: *Malnutrition, Behavior and Social Organization*, ed. L. S. Greene & F. E. Johnston, pp. 55–94. New York: Academic Press.

Greksa, L. P. (1990). Developmental responses to high altitude hypoxia in Bolivian children of European ancestry: Test of the developmental adaptation hypothesis. *American Journal of Human Biology* **2**, 603–12.

Greksa, L. P. (1991) Human physiological adaptation to high altitude environments. In: *Applications of Biological Anthropology to Human Affairs*, ed. C. G. N. Mascie-Taylor & G. W. Lasker, pp. 117–42. Cambridge University Press.

Haas, J. D. (1976). Infant growth and development. In: *Man in the Andes: a multidisciplinary study of high-altitude Quechua natives*, ed. P. T. Baker & M. A. Little, pp. 161–9. Stroudsburg, Pennsylvania: Dowden, Hutchinson & Ross.

Haas, J. D. (1980). Maternal adaptation and fetal growth at high altitude in Bolivia. In: *Social and Biological Predictors of Nutritional Status, Physical Growth, and Neurological Development*, ed. L. S. Greene & F. E. Johnston, pp. 257–90. New York: Academic Press.

Harrison, G. A. (1966). Human adaptability with reference to the IBP proposals for high altitude research. In: *The Biology of Human Adaptability*, ed. P. T. Baker & J. S. Weiner, pp. 509–19. Oxford: Clarendon Press.

Harrison, G. A. (1988). Human genetics and variation. In: *Human Biology*, 3rd edn, ed. G. A. Harrison, J. M. Tanner, D. R. Pilbeam & P. T. Baker, pp. 147–335. Oxford University Press.

Harrison, G. A., Tanner, J. M., Pilbeam, D. R. & Baker, P. T. (1988). *Human Biology*, 3rd edn. Oxford University Press.

Himes, J. H. (ed.) (1991). *Anthropometric Assessment of Nutritional Status*. New York: Wiley–Liss.

Johnston, F. E. (1982). *Physical Anthropology*. Dubuque, Iowa: Wm. C. Brown.

Johnston, F. E., Weiner, H., Thissen, D. & McVean, R. B. (1976). Hereditary and environmental determinants of growth in height in a longitudinal sample of children and youth of Guatemalan and European ancestry. *American Journal of Physical Anthropology* **44**, 469–75.

Kaplan, B. (1954). Environment and human plasticity. *American Anthropologist* **56**, 780–99.

Kelso, A. J. (1970). *Physical Anthropology*. Philadelphia: J. B. Lippincott.

Kuhn, T. S. (1962). *The Structure of Scientific Revolutions*. Chicago: University of Chicago Press.

Lasker, G. W. (1969). Human biological adaptability. *Science* **166**, 1480–6.

Little, M. A. (1982). The development of ideas on human ecology and adaptation. In: *A History of American Physical Anthropology, 1930–1980*, ed. F. Spencer, pp. 405–33. New York: Academic Press.

Little, M. A. & Morren, G. E. B. Jr. (1976). *Ecology, Energetics, and Human Variability*. Dubuque, Iowa: Wm. C. Brown.

Martorell, R., Dellgaso, H. L., Valverde, V. & Klein, R. E. (1981). Maternal stature, fertility and infant mortality. *Human Biology* **53**, 303–12.

Mazess, R. B. (1975). Biological adaptation: aptitudes and acclimatization. In: *Biosocial Interrelations in Population Adaptation*, ed. E. S. Watts, F. E. Johnston & G W. Lasker, pp. 9–18. The Hague: Mouton Publishers.

McElroy, A. & Townsend, P. K. (1979). *Medical Anthropology in Ecological Perspective*. North Scituate, Massachusetts: Duxbury Press.

Moore, L. G. (1990). Maternal O_2 transport and fetal growth in Colorado, Peru, and Tibet high-altitude residents. *American Journal of Human Biology* **2**, 627–37.

Mueller, W. H. (1979). Fertility and physique in a malnourished population. *Human Biology* **51**, 153–66.

Mueller, W. H., Yen, F., Rothhammer, F. & Schull, W. J. (1978). A multinational Andean genetic and health program. VI. Physiological measurements of lung function in an hypoxic environment. *Human Biology* **50**, 489–513.

Newman, M. T. (1953). The application of ecological rules to the racial anthropology of the aboriginal New World. *American Anthropologist* **55**, 311–27.

Newman, R. W. & Munro, E. H. (1955). The relation of climate and body size in U.S. males. *American Journal of Physical Anthropology* **13**, 1–17.

Poirier, F. E., Stini, W. A. & Wreden, K. B. (1990). *In Search of Ourselves*, 4th edn. Englewood Cliffs, New Jersey: Prentice-Hall.

Prosser, C. L. (1964). Perspectives of adaptation: theoretical aspects. In: *Handbook of Physiology*, Vol. 4, *Adaptation to the Environment*, ed. D. B. Dill, E. F. Adolph & C. G. Wilber, pp. 11–25. Washington, D.C.: American Physiological Society.

Relethford, J. H. (1994). *The Human Species*, 2nd edn. Mountain View, California: Mayfield.

Roberts, D. F. (1953). Body weight, race and climate. *American Journal of Physical Anthropology* **4**, 533–58.

Roberts, D. F. (1993). Human biology in international research programmes. *Annals of Human Biology* **20**, 1–11.

Scheffler, I. (1962). *Science and Subjectivity*. Indianapolis: Bobbs-Merrill.

Schell, L. M. (1992). Pollution and human growth: lead, noise, polychlorinated biphenyl compounds and toxic waste. In: *Applications of Biological Anthropology to Human Affairs*, ed. C. G. N. Mascie-Taylor & G. W. Lasker, pp. 85–116. Cambridge University Press.

Schell, L. & Ando, Y. (1992). Postnatal growth of children in relation to noise from Osaka International Airport. *Journal of Sound and Vibration* **151**, 371–82.

Scrimshaw, N. S. & Young, V. R. (1989). Adaptation to low protein and energy intakes. *Human Organization* **48**, 20–30.

Seckler, D. (1982). Small but healthy: A basic hypothesis in the theory, measurement, and policy of malnutrition. In: *Newer Concepts of Nutrition and their Implications for Policy*, ed. P. V. Sukhatme, pp. 127–37. Pune, India: Maharashta Association for the Cultivation of Science Research Institute.

Shapiro, S., McCormick, M. C., Starfield, B. H., Krischer, J. P. & Bross, D. (1983). Relevance of correlates of infant deaths for significant morbidity at 12 months of age. *American Journal of Obstetrics and Gynecology* **136**, 363–73.

Slobodkin, L. B. (1968). Toward a predictive theory of evolution. In: *Population Biology and Evolution*, ed. R. C. Lewontin, pp. 187–205. Syracuse: Syracuse University Press.

Stein, P. L. & Rowe, B. M. (1989). *Physical Anthropology*, 4th edn. New York: McGraw-Hill.

Stini, W. A. (1975). *Ecology and Human Adaptation*. Dubuque, Iowa: Wm. C. Brown.

Stinson, S. (1992). Nutritional adaptation. *Annual Reviews in Anthropology* **21**, 143–70.

Tanner, J. M. (1966). Growth and physique in different populations of mankind. In: *The Biology of Human Adaptability*, ed. P. T. Baker & J. S. Weiner, pp. 45–66. Oxford: Clarendon Press.

Tanner, J. M. (1979). A concise history of growth studies from Buffon to Boas. In: *Human Growth* (1st edn.), *Neurobiology and Nutrition*, ed. F. Falkner & J. M.

Tanner, pp. 515–93. New York: Plenum Press.

Tanner, J. M. (1986). Growth as a mirror of the condition of society: Secular trends and class distinctions. In: *Human Growth: A Multidisciplinary Review*, ed. A. Demirjian, pp. 3–34. New York: Plenum.

Thomas, R. B. (1975). The ecology of work. In: *Physiological Anthropology*, ed. A. Damon, pp. 59–79. Oxford University Press.

Waaler, H. T. (1984). Height, weight, and mortality: the Norwegian experience. *Acta Medica Scandinavica*, **679** (suppl.), 1–56.

Weiner, J. S. (1966). Major problems in human population biology. In: *The Biology of Human Adaptability*, ed. P. T. Baker & J. S. Weiner, pp. 1–24. Oxford: Clarendon Press.

Weiss, K. & Mann, A. (1991). *Human Biology and Behavior*, 5th edn. Boston: Little, Brown & Co.

Index